Robert D. Bush

Service
PARTS
Management

Joseph D. Patton, Jr.

Service
PARTS
Management

Instrument Society of America

SERVICE PARTS MANAGEMENT

Printed in the United States of America

In preparing this work, the author and publisher have not investigated or considered patents which may apply to the subject matter hereof. It is the responsibility of the readers and users of the subject matter to protect themselves against liability for infringement of patents. The information contained herein is of a general educational nature. Accordingly, the author and publisher assume no responsibility and disclaim all liability of any kind, however arising, as a result of using the subject matter of this work.

The equipment referenced in this work has been selected by the author as examples of the technology. No endorsement of any product is intended by the author or publisher. In all instances, the manufacturer's procedures should prevail regarding the use of specific equipment. No representation, expressed or implied, is made with regard to the availability of any equipment, process, formula, or other procedures contained herein.

Instrument Society of America
67 Alexander Drive
P.O. Box 12277
Research Triangle Park, NC 27709

Library of Congress Cataloging in Publication Data

Patton, Joseph D.
    Service parts management.

    Bibliography: p.
    Includes index.
    1. Inventory control.   2. Machine parts.   I. Title.
TS160.P37 1984   658.7′8   84-19163
ISBN 0-87664-811-1

Design and book production
by Publishers Creative Services Inc., New York

Contents

13. PRICING   203

14. MODELS   211

15. REPAIRABLES   227

16. COMPUTERIZATION   241

17. PERSONNEL AND ORGANIZATION   257

List of Figures

List of Tables

Preface

OBJECTIVES

HE OBJECTIVES OF THIS text are to:

1. Develop understanding of the terms, principles, and practices important to effective management of service consumables, durable equipment, materials, and parts.

2. Create awareness of the differences between the management of parts necessary for maintenance service and those used in support of manufacturing.

3. Improve knowledge of the considerations and trade-offs between labor and parts, with the ability to apply life-cycle cost and profit analysis.

4. Provide a reference book with practical guidelines that can be of immediate help to a person who is told, "I want you to get in there and straighten out our service parts!"

OVERVIEW

The objective of service parts management is to satisfy customers by providing the necessary parts, at the user's location, for a reasonable price, in usable condition, at the proper time.

This can be referred to as the five rights of parts: item, place, price, quality, and time.

A major trade-off of the service business is parts versus labor. As labor costs increase, the trade-off will shift in favor of using more parts and less labor. This is also stimulated by customers' needs for high equipment uptime that can best be obtained by a quick change of complete modules rather than extensive on-site repair with piece parts.

Financial pressures on parts are also increasing. In the "Fortune 500" largest U. S. industrial corporations alone, an estimated $115 billion is invested in service parts. Inflation means that generally the next parts order will cost more than the last one did. Carrying costs of inventory for space, security, environment, record keeping, computer systems operations, insurance, obsolescence, personnel, and federal and state taxes, and the cost of borrowing or alternative uses of money, all direct increased attention to service parts.

Availability of materials such as petrochemical-based plastics, gold for electronic contacts, and tungsten for springs is a concern. Lead times for electronic circuit boards, bearings, and similar items that used to be available "off the shelf" have stretched out to many months, even into years. The capability of small businesses to produce custom parts rapidly has been hindered by the economic factors that limit investment in inventories and increase the risks and costs all the way from the source of raw material to the finished components.

The service parts function in many organizations needs to be accorded higher status and to have more importance placed upon it. In the past, personnel were sometimes exiled to the parts function when they were unable to do a good job elsewhere. Fortunately that is changing. However, parts management personnel are often paid very low wages to manage millions of dollars worth of inventory. The parts business should receive attention not only when trouble occurs, but also during the routine times that probably indicate good management. Service parts today is frequently organized as a profit center, and some businesses make as much profit from service parts as they do from the sale of finished goods. Good parts support should be recognized and praised!

Parts is still a people business. Computer power enables us to process information accurately and rapidly. Innovations in the implementation of our mental processes, many of which are outlined in this book, provide more knowledge. The necessary additional ingredient is motivated managers who can apply this knowledge successfully to specific situations.

A great need exists for practical information on the principles of service parts management and how to practice them. Billions of dollars are invested in inventories of equipment, materials, and parts. Hundreds of thousands of people hold such jobs as materials forecaster, inventory control specialist, stockkeeper, receiving clerk, distribution manager, expeditor, provisioning analyst, picker, packer, manager of material, service planning specialist, and logistician. The rapid transportation business, including such companies as Emery Air, Federal Express, and United Parcel Service, has been built around the need to move service parts rapidly. "I don't have the right part!" is probably heard from more field service and plant maintenance personnel than any other complaint. On the other hand, many stockrooms and warehouses contain items that have not been touched for a long time. In fact, one survey of 230 industrial dealers showed that 42 percent of all the part numbers stocked had not moved in over two years!

Service parts are related to manufacturing production parts. The problems and opportunities are also kin to the challenges of wholesalers and retailers in keeping their shelves stocked. Service parts, however, do have many significant differences, which are a focus of this book.

Professional associations whose members should be interested in this information include:

American Institute of Plant Engineers (AIPE)

American Production and Inventory Control Society (APICS)

American Society for Quality Control (ASQC)

Association of Field Service Managers (AFSM)

Commercial Food Equipment Service Agencies (CFESA)

Computer and Business Equipment Manufacturers Association (CBEMA)

Equipment Maintenance Council (EMC)

Farm and Industrial Equipment Institute (FIEI)

Industrial Maintenance Institute (IMI)

Institute of Electrical and Electronics Engineers (IEEE)

Institute of Industrial Engineers (IIE)

Institute of Industrial Launderers (IIL)

National Association of Food Equipment Manufacturers (NAFEM)

National Association of Service Managers (NASM)

National Council of Physical Distribution Management (NCPDM)

National Farm Power and Equipment Dealers Association (NFPEDA)

Service Council of the Forestery Management Council (FMC)

Society of Logistics Engineers (SOLE)

Textile Rental Services Association

A distinction will be made between "equipments," which are considered durable and are individually marked with a unique serial number identifier; "parts," which are stockkeeping units in which many items may be (in fact, should be) identical; and "materials" such as cleaning supplies, lubricants, and paint. All three classes are included in our coverage with note made of important differences as they affect the topic.

This material has been used to conduct many public workshops for leading universities and professional associations, and in internal training for many corporations and governmental agencies. During the period of economic recession, when this text was prepared, workshops on "Service Parts Management" were usually well attended, in contrast to people-related workshops. International aspects are considered because the problems are generally the same in the United States as they are in Australia, Israel, or Zambia. Improving service parts management will have a definite positive impact on commerce, government, and industry—with benefit to people around the world.

ACKNOWLEDGMENTS

The author's technique for converting thoughts into text is first to prepare an outline of chapters and major topics. Then most of the words are dictated into an electronic portable tape recorder. The drive from home to office allows about 20 minutes of dictation most mornings. The evening drive home is more often spent listening to soothing music instead of dictating. For this book Beverly Phillips transcribed the taped words onto the computer line editor and integrated the illustrations. After initial editing, the manuscript went to the publisher for circulation to experienced managers for their comments and additional ideas. Then the editors and illustrators plied their craft. The marked manuscript was read by the author and final decisions rendered. The manuscript next appeared in type for proofreading to ensure that all the words were in place and spelled correctly. Then page proofs presented a final opportunity to assure quality content. The printer, binder, and publisher packaged the final product, resulting in finished books about 18 months after the first outline was prepared. This book has passed through concept, design, development, and production phases as do most durable goods and the pieces have finally come together to form *Service Parts Management.*

Sidney and Raymond Solomon of Publishers Creative Services deserve credit for persevering through this fourth book with the author. The author's fun is over when the words are dictated. Fortunately, the production process that is drudgery to the busy author is what "turns them on."

Special thanks are due the professional service parts managers and consultants who reviewed the initial manuscript and offered constructive suggestions for improvement. They include:

<table>
<tr><td>

Keith P. Chase
    Consultant
    Patton Consultants, Inc.

David C. Davis
    Customer Equipment
      Service Division
    Eastman Kodak Corporation

</td><td>

Herbert C. Feldmann
    Vice President—
      Management Systems
    Patton Consultants, Inc.

Leon Jackson
    Consultant
    Patton Consultants, Inc.

</td></tr>
</table>

Larry G. DeVries
Manager of Inventory
Control
Control Data Corporation

Willem A. Dieleman
Manager, Service Parts
Center
Service Parts Center/3M
Company

Terry Sexton
Service Parts Manager
Steiger Tractor, Inc.

George Smith
President
Marshall Institute

John Woltman
Advanced Operations
Research Planner
Service Parts Center/3M
Company

1

Terms
and
Definitions

THE SERVICE PROFESSION has developed some unique terms and definitions to describe parts-related activities. Most of the words used to describe the business of managing service parts are derived from their common use in other areas and logical dictionary references. Many readers will wish to skim through the following pages and move rapidly on to the next chapter. Others, new to the realm of EOQ, ROP, and SKUs, will want to use the following terms and definitions as a reference base. If the meaning of a word used in later chapters is not clear, then review of this chapter should provide clarification.

ABC inventory policy Collection of prioritizing practices to give varied levels of attention to different classes of inventories.

Allocations Apportionments made for specific purposes to two or more persons or things or locations.

Availability With reference to parts, the fact that part is on hand and in usable condition. Product availability is the probability that the item will work properly under stated conditions. Inherent availability (A_i) concerns failures only. Achieved availability (A_a) includes preventive maintenance and planned downtime. Operational availability (A_o) considers total downtime, including administrative and logistics times, so

$$A_o = \frac{\text{uptime}}{\text{total time.}}$$

Capital Durable items with long life or high value that necessitates asset control and depreciation under tax guidelines, rather than being expensed.

Carrying costs Expense of handling, space, information, insurance, special conditions, obsolescence, personnel, and the cost of capital or alternate use of funds to keep parts in inventory; also called holding costs.

Chaining Procedure whereby on orders for part numbers that have been replaced the oldest number will be checked and the order filled with it if possible. If those parts no longer exist, the link goes to the stock of parts under the "replaced by" number.

Commodity code Classification of parts by group and class according to their material content or type for consolidation of procurement, storage, and use.

Confidence Degree of certainty that something will happen. For example, a low confidence of replenishment means parts probably will not be rapidly replaced, and is the main reason field service personnel retain excess parts.

Configuration Physical and functional characteristics of systems and items; the specific parts used to construct a machine; the shape or makeup of a thing at a given time.

Consumables Supplies such as fuel, lubricants, paper, printer ribbons, cleaning materials, and forms that are exhausted during use in operation and maintenance.

Contingency Alternate actions that can be taken if the main actions do not work.

Cost of sales Cost of transfer from the supplier plus the cost for operating the service parts function.

Critical   Category of items that are very important to performance and are more vital to operation than are noncritical items; the "significant few," synonymous with essential.

Dead stocks   Items for which no demand has occurred over a specific period of time.

Demand   Requests and orders for an item. Demands become issues only when a requested part is given from stock.

Direct costs   Any expenses that can be associated with a specific product, operation, or service.

Disposal   Act of getting rid of surplus property under proper authorization. Typical means of disposal are abandonment, destruction, donation, recycling, sale, scrap, sell back to vendor/manufacturer, and transfer.

Echelon stock   Intermediate stock levels in a logistics network between the central supply point and the point of maintenance.

Economic order quantity (EOQ)   Amount of an item that should be ordered at one time to get the lowest possible combination of inventory-carrying and order/production costs.

Engineering change   Any design change that will require revision to specifications, drawings, documents, or configurations.

Environment   Aggregate of all conditions and influences on parts and equipment. Typical considerations are physical location, operating characteristics of nearby equipment, actions of people, and conditions of temperature, humidity, salt spray, acceleration, shock, vibration, radiation, and contaminants in the surrounding air or liquid.

Error type one/alpha/producer   Mistake made when items thought to be defective are, in fact, good.

Error type three/Charlie/problem   Mistake made when the problem is not correctly identified.

Error type two/beta/consumer   Mistake made when items are thought to be good but are, in fact, defective. The acronym BAB (beta accept bad) should help in remembering the error types.

Essentiality Importance of an item to performance of the mission.

Excess Items declared to be more than projected for current needs. Several categories of excess may be used to segregate materials for attention. Typical categories are:

1. Excess with forecast need.
2. Excess with no forecast need and no use in last 12 months.
3. Excess with no forecast need and no use in last 12–24 months.
4. Excess with no forecast need and no use in last 25+ months.

Expediting Special efforts to accelerate a process. An expediter coordinates and assures adequate supplies of materials and equipment.

Expense Those items that are directly charged as a cost of doing business. They generally have a short, nondurable life. Most nonrepairable service parts are expensed when installed on equipment.

Expensed inventory Parts written off as a "cost of sales." Material transferred from ledger inventory to expensed inventory is to be used within 12 months. Most often the parts remain under the control of field service personnel in order to support maintenance under contract.

Exponential smoothing Mathematical technique that weights new data differently than old; also called a geometric moving average.

Failure Inability to perform the function within specified limits.

Failure modes, effects, and criticality analysis (FMECA) Reliability analysis of what items are expected to fail, the consequences of failure, and how important they are to system performance.

Fill rate Service level of a specific stock point. An 85 percent fill rate means that if 100 parts are requested, then 85 of them are available and issued.

Finite replenishment rate Specific, usually uniform, rate of replenishment; for example, 100 units per month.

First in-first out (FIFO) Use the oldest item in inventory next. FIFO accounting values each item used at the cost of the oldest item in inventory. Contrasts with LIFO (last in-first out).

First in-still here (FISH) Result of poor forecasting and management. Any FISH parts should be removed from stock unless they are insurance-class parts because they are essential and have a long lead time.

Fractionation Supply process whereby stocked items are classified as to relative rate of issue, cost, or other significant factors.

Frequency Count of occurrences during each time period or event. A typical frequency chart for inventory plots demand versus days.

Gross profit Amount of profit as a percent of sales revenues.

Histogram Graphic plot of a frequency distribution by means of rectangles whose widths normally represent class intervals and whose heights represent corresponding frequencies.

Hold for disposition stock Defective material held at a stock location pending removal for repair or for scrap.

Incremental analysis Analysis of the benefit or cost related to change; also called marginal analysis.

Initial demand and supply Earliest period of field introduction in which pipeline, shelf, training, and failure quantities must be taken into consideration.

Insurance items Parts and materials that are not used often enough to meet detailed stock accounting criteria, but are stocked because of their essentiality or the lead time involved in procuring replacements; similar to safety stocks, except on low-use parts.

Integrated logistics support (ILS) Composite of the elements necessary to assure the effective and economical sustaining of a system or equipment, at all levels of maintenance, throughout its programmed life cycle. It is characterized by the harmony and coherence obtained between each of its elements and levels of maintenance.

Interchangeable Parts with different configurations and numbers that may be substituted for another part, usually without any modification or different performance, since they have the same form, fit, and function.

Inventory Physical count of all items on hand by number, weight, length, or other measurement; also any items held in anticipation of future use.

Inventory control That phase or function of logistics that includes management, cataloging, requirements determination, procurement, inspection, storage, distribution, overhaul, and disposal of material.

Investment stock Materials held for speculation in the hope that their value will increase owing to inflation or other economic conditions.

Item Generic term used to identify a specific entity. Items may be parts, components assemblies, subassemblies, accessories, groups, parents, components, equipments, or attachments.

Last in-first out (LIFO) Use newest inventory next. LIFO accounting values each item used at the cost of the last item added to inventory. Contrasts with FIFO (first in-first out).

Lead time Allowance made for that amount of time estimated or actually required to accomplish a specific task such as acquiring a part.

Ledger inventory Items carried on the corporate financial balance sheet as material valued at cost.

Level of repair (LOR) Locations and facilities at which items are to be repaired. Typical levels are operator, field technician, bench, and factory.

Life cycle Series of phases or events that constitute the total existence of anything. The entire "womb-to-tomb" scenario of a product from the time concept planning is started until it is finally discarded.

Life-cycle costs All costs associated with the system life cycle, including research and development, production, operation, maintenance, and termination.

Logistics   Art and science of management and engineering, and the technical activities concerned with requirements, design, and supplying and maintaining resources to support objectives, plans, and operations.

Lower of cost or market   Value placed on inventory that uses the lower of what was paid for an item or what it could be sold for now.

Maintenance concept   Statements and illustrations that define the theoretical means of maintaining equipment. It relates tasks, techniques, technology, tools, parts, and people.

Marginal analysis   Synonymous with incremental analysis.

Material   Metal, wood, lubricants, cloth, and other hard goods generally purchased in bulk and used in smaller, variable quantities; also all items used or needed in any business, industry, or operation as distinguished from personnel.

Materiel   Military term that covers all items necessary for the equipment, maintenance, operations, and support of military activities, whether for administrative or combat use, and excluding ships and aircraft.

Mean, arithmetic   Average of a series of values; specifically, their sum divided by the number of items in the series. Sometimes called simply "mean," and symbolized by $\bar{X}$ ("X bar").

Mean downtime (MDT)   Average time a system cannot perform its mission, including response, active maintenance, and supply and administrative time.

Mean logistics delay time (MLDT)   Downtime while necessary replacement parts, supplies, tools, or data are being obtained.

Mean time between failures (MTBF)   Average time/distance/events during which a part or equipment performs between breakdowns.

Mean time between replacement (MTBR)   Average time/distance/events between when a part is installed on equipment and when it is removed for any reason. Typically this is MTBF less the interval between scheduled preventive maintenance replacements.

Median Midpoint of a series of quantities or values; specifically, the quantity or value of that item which is so positioned in the series when arranged in order of numeric quantity or value that there are an equal number of values of greater magnitude and of lesser magnitude.

Mode Most frequently occurring value or type in a series.

Model Simulation of an event or process or product physically, mathematically, or verbally.

Monte Carlo analysis Simulation in which occurrences have a probability of their occurring proportional to expectations in the "real world."

Moving average Mathematical method of considering recent events; for example, averaging the last three months' shipments or the cost of the last ten items added to inventory.

Network fill rate Percent of orders filled completely at a local stock point plus additional stock points in the distribution network.

Next higher assembly Component level of hierarchy immediately superior to the item in question; "parent" item in manufacturing.

Nomographs Charts that relate mathematical models and provide solutions by connecting scales with straight lines.

Nonrepairable Parts or items that are discarded upon failure for technical or economic reasons.

Normal distribution Statistical distribution commonly referred to as a "bell curve." The mean, mode, and median are the same in a normal distribution.

Not-operational-ready-spares (NORS) Equipment that is not available for use because of a lack of parts.

Obsolescence Decrease in value or use of items that have been superseded by superior items.

Obsolete Designation of an item for which there is no replacement. The part has probably become unnecessary as a result of a design change.

Operating days (O days) Number of days on which production activities take place. There are normally about 250 operating days in a year.

Opportunity cost Value of either gaining or failing to gain projected business. For example, if a customer requests a part that is not available, the customer may well go elsewhere for that part and perhaps also for other future purchases. The opportunity cost of that part would be the potential profit from that customer's business. This is also termed "lost sales" or "lost sales cost."

Optimal replenishment model Mathematical process of determining what quantity to order and when, by using a fixed order point or a fixed order interval.

Order point (P) Quantity of parts at which an order will be placed when usage reduces stock to that level; also called reorder point (ROP).

Order quantity (Q) Number of items demanded. The economic order quantity (EOQ), also called minimum cost quantity, is a specific number; but the actual order quantity may vary as a result of cost, transportation, discounts, or extraordinary demand.

P model Mathematical logic for determining quantities on the basis of fixed order point (P).

Packaging Use of protective wrappings, cushioning, containers, and complete identification marking, up to but not including the exterior shipping container.

Packing Application or use of exterior shipping containers or other shipping media (such as pallets), and the assembling of items or packages thereof, together with the necessary blocking, bracing, or cushioning; weatherproofing; exterior strapping; and marking of the shipping container.

Pareto's principle Critical few, often about 20 percent, of parts or people or users that should receive attention before the insignificant many, which are usually about 80

Part numbers Unique identifying numbers and letters that denote each specific part configuration; also called stock numbers or item numbers.

Perpetual record Constantly updated records that will identify the status of parts at any present point in time.

Physical distribution Movement, control, protection, and storage of materials. Activities included are freight, transportation, warehousing, material handling, protection, packaging, market forecasting, and customer service.

Pipeline Channel of support by means of which material or personnel flow from sources of procurement to their point of use, or from point of failure to repair point.

Pipeline stock Material moving to or from one point to another, usually from the source to the point of issue, or defective material moving from the point of failure to the repair point.

Poisson Frequency distribution that is a good approximation of the binomial distribution as the number of trials increases and the probability of success in a single trial is small; often used to describe demand queueing.

Predictive maintenance Subset of preventive maintenance that uses nondestructive testing such as spectral oil analysis, vibration evaluation, and ultrasonics with statistics and probabilities to predict when and what maintenance should be done to prevent failures.

Preventive maintenance (PM) Actions performed in an attempt to keep an item in a specified operating condition by means of systematic inspection, detection, and prevention of incipient failure.

Priority order classification Order of importance and shipping sequence in order-processing systems. Typically, A = system down; B = system degraded; C = equipment degraded; K = kit part replenish; R = routine. Ship sequence is A, B, C, K, R.

Procurement Process of obtaining persons, services, supplies, facilities, materials, or equipment. It may include the function of design, standards determination, specification development, selection of suppliers, financing, contract administration, and other related functions.

Product code Designation used to extract part groups from a master file according to product. For example, "ABN" could be A = product line, B = equipment within product, N = basic equipment, option, or kit.

Provisioning Process of determining and selecting the varieties and quantities of repair parts, spares, special tools, and test and support equipment that should be procured and stocked to sustain and maintain equipment for specified periods of time. It includes identification of items of supply; establishing data for catalogs, technical manuals, and allowance lists; and providing instructions and schedules for delivery of provisioned items.

Purchase order costs Expenses associated with placing a purchase order, includes gathering correct information, typing the requisition, gaining approvals, getting quotes, selecting the vendor, placing the purchase order, and receiving inspection.

Q model Type of inventory model that operates on a fixed order quantity (Q).

Queuing Pattern of demand placed on any activity, such as an accumulation of work at a work station.

Ready for issue Category of good stock available for immediate shipment.

Records Data, facts, and information. The media may range from a small black book in the worker's pocket to an electronic computer file. A computerized part record contains all fields of data about a part.

Reliability Probability that any item will perform its intended function without failure for a specified time under specified conditions.

Reorder point (ROP) Minimum quantity, established by economic calculation and management direction, that triggers the ordering of more items.

Repair Restoration or replacement of parts or components as necessitated by wear, tear, damage, or failure; to return the facility, equipment, or part to efficient operating condition.

Repair code Yes/no designation of whether or not a part may be repaired.

Repair component parts Parts below the field maintenance level that are needed to repair failed spare parts at the repair point.

Repair parts Individual parts or assemblies required for the maintenance or repair of equipment, systems, or spares. Such repair parts may also be repairable or nonrepairable assemblies, or one-piece items. Consumable supplies used in maintenance or repair, such as wiping rags, solvents, and lubricants, are not considered repair parts. Repair parts are also service parts.

Repair point Location at which defective repairables are repaired and returned to a ready-for-issue condition.

Repairables Parts or items that are technically and economically repairable. A repairable part, upon becoming defective, is subject to return to the repair point for repair action.

Replaced by Term applied to a part number that has been superseded by another part.

Retrofits Part, assembly, or kit that will replace similar components originally installed on equipment. Retrofits are generally performed on a machine to correct a deficiency or to improve performance.

Revenue Money received for supplying parts, labor, equipment, or other products and services.

Safety stock Quantity of an item, in addition to the normal level of supply, required to be on hand to permit continuing operation with a specific level of confidence if resupply is late or demand suddenly increases.

Scenario   Description in words and/or graphics that describes how an item will be used, or how a situation can develop.

Scheduled maintenance (SM)   Subset of preventive maintenance based on fixed intervals of time or use or events.

Scrap   Items discarded in so far as original use is concerned, which have no reasonable prospect of value except for the recovery of their basic material content.

Seasonal trends   Variations in demand that fluctuate with the seasons of a year. The influencing factors may be weather, vacations, holidays, growing cycles, or similar events that are repeated year after year.

Secondary failures   Malfunctions that are caused by the failure of another item.

Service level   Frequency, usually expressed as a percentage, with which a demand can be filled through a particular stock echelon. A 95 percent level of service means that 95 out of 100 demands are properly issued. If viewed from the end customer or service technician perspective, the service level is the percent of parts received out of those requested, from any and all levels of the support system.

Service parts   Parts used for the repair and/or maintenance of an assembled product (APICS).

Setup costs   All expenses necessary to prepare for manufacturing an ordered part. Setup costs are usually assumed to be constant regardless of the quantities produced, so the cost per unit will decrease as order size increases. Setup costs may increase substantially in later stages of the life cycle after routine production is stopped, since additional retooling, material, information, training, and personnel relearning is required.

Shelf life   Period of time during which an item can remain unused in proper storage without significant deterioration.

Shipment   Item or group of items from one place, released to a carrier for transportation to a single destination.

Simulation   Mathematical process that attempts to model what will happen in the real world. An inventory simulation percent.

is usually run on a computer with probabilities of demand and supply injected according to typical patterns.

Software Efforts, plans, and documentation to support or sustain projects, operations, equipment, and items; including such things as engineering and design, technical data, plans, schedules, and computer programs. Software excludes physical parts, materials, equipment, and tools; and contrasts with hardware.

Spare parts (spares) Components, assemblies, and equipment that are completely interchangeable with like items installed or in use, which are used, or can be used, to replace items removed during maintenance and overhaul. These parts used for maintenance of an assembled product are also called field replaceable units (FRUs). "Service parts" is a more desirable name because "spare" has the connotation of being extra.

Specifications Documents that clearly and accurately describe the essential requirements for materials, items, equipment, systems, or services, including the procedures by which it will be determined that the requirements have been met. Such documents may include performance, support, preservation, packaging, packing, and marking requirements.

Standard deviation Measure of average dispersion (departure from the mean) of numbers, computed as the square root of the average of the squares of the difference between numbers and their arithmetic mean.

Standard item Part, component, material, subassembly, assembly, or equipment that is identified or described accurately by a standard document or drawing.

Standardization Process of establishing the greatest practical uniformity of items and of practices to assure the minimum feasible variety of such items and practices, and effect optimum interchangeability.

Standards Established or accepted rules, models, and criteria by which the degree of user satisfaction of a product or an act is determined, or against which comparisons are made. A document that establishes engineering and technical limitations for items, materials, processes, methods, designs, and practices is therefore a standard.

Statistics   Collecting, classifying, summarizing, and interpreting of numeric facts by other than accounting methods.

Stock number   Number assigned by the stocking organization to each group of articles or material, which are then treated as if identical within the using supply system; also called part number or item number or part identifier.

Stock out   Indicates that all quantities of a part normally on hand have been used, so that the items are not presently available. Demand for a nonstock part is usually treated as a separate situation.

Stock-point fill rate   Percent of orders filled completely on the first pass.

Stockkeeping unit (SKU)   Reference that designates each SKU from others according to shape, size, color, fragrance, strength, reliability, or other characteristics. An SKU inventory means the stock of an individually described SKU that may contain any quantity of units.

Stockkeeping unit location (SKUL)   Number of units inventoried at one facility. One or more SKUs make up a product line and the sum of all SKUs make up a total inventory for a company.

Supply   Procurement, storage, and distribution of items.

System   Assemblage of correlated hardware, software, methods, procedures, and people, or any combination of these, all arranged or ordered toward a common objective.

Throwaway maintenance   Maintenance performed by discarding used parts rather than attempting to repair them.

Trial products   Test items usually built in small quantities for research and development, experimentation, development, testing, and demonstration prior to production.

Turnaround time   Interval between the time a repairable item is removed from use and the time it is again available in serviceable condition.

Turnover Measurement on either numbers of parts or on monetary value that evaluates how often a part is demanded versus the average number kept in inventory. For example, if two widgets are kept in inventory and eight are used each year, then the turnover is $8/2 = 4X$ per year. In monetary terms turnover is cost of inventory sold/average cost of inventory carried.

Uncertainty Lack of definite knowledge that ranges between complete certainty and a total lack of knowledge. Variability of demand and supply creates uncertainty.

Universal Product Code (UPC) Rectangular array of bars and numbers that can be scanned to send information to a computer that processes the data for pricing and inventory control.

Unscheduled maintenance (UM) Corrective maintenance (CM), emergency maintenance (EM), or repairs to restore a failed item to usable condition; contrasts with scheduled maintenance.

Upgradability Yes or no code that indicates feasibility of modifying the previous revision of a part to the current revision.

Usage Quantity of items consumed or necessary for product support. Usage is generally greater than the technical failure rate.

Warranty Guarantee that an item will perform as specified for at least a specified time, or will be repaired or replaced at no cost to the user.

Wear out Deterioration as a result of age, corrosion, temperature, or friction that generally increases the failure rate over time.

Objectives
for
Service Parts

EOPLE SOMETIMES USE THE word "spare" in referring to parts. That is a problem symptom. If the word is defined in the sense of the "spare" tire for a car—that it is there for emergency backup insurance use—that usage is fine. Often, however, spare means excess to any reasonable requirements. Stockkeepers are by nature hoarders. Most people in the service business have a caché of parts that they expect to use eventually. Many parts turn over less than once a year, and wiping a finger over them often will show thick layers of residue from lack of use. Also, some of the parts may be for equipment that has long since been scrapped. On the other hand, a part that is requested often is not available. That absence may cause production loss, thus promoting higher costs, stress, and lack of confidence by users in the service parts system.

It is necessary for every organization to have objectives and goals. Such goals should be written, understandable, measurable, challenging, and achievable. The following are goals that have been adopted by many organizations and discussed with thousands of service personnel in workshops and presentations. Most organizations can meet a few of the goals; some meet all of them. The purpose of stating the goals is to set our sights high and to focus future learning on the practical question of how to meet these goals in your own situation. If your organization cannot meet a specific goal, then identify how good (or bad) you are today, and negotiate an improved goal as the target.

PERFORMANCE GOALS FOR SERVICE PARTS

1. QUALITY

Completeness

A central parts facility shall be able to fill at least 95 percent (19/20) of all orders the same day. This includes issuing at least 95 percent of requested line items.

Field personnel shall have access to parts that fill at least 85 percent of their parts needs within 15 minutes.

End-user customers and field personnel shall receive at least 95 percent of needed essentiality 1 or 2 (critical equipment down) parts within four hours, and 98 percent of the parts within 24 hours.

Accuracy

Orders received by customers and field personnel shall be at least 98 percent correct.

Packing lists shall accurately indicate 99 percent of the items shipped.

Defects attributable to poor quality or damage shall not account for more than 1 percent of parts received.

2. TIME

Emergency

All orders received by noon shall be shipped the same day by the fastest method.

Regular

All orders shall be entered the same day as received.

Ninety-five percent of in-stock items shall be selected and shipped by noon of the next workday.

One hundred percent of in-stock items shall be selected and shipped by noon of the second workday.

Out of Stock or Nonstock

Ninety percent shall be scheduled within two workdays.

One hundred percent shall be scheduled within four workdays.

Ninety percent shall be shipped within ten workdays.

Ninety-five percent shall be shipped within 20 workdays.

All back orders for down equipment parts received by noon shall be shipped the same day. All other back-ordered parts due for shipment shall be shipped by noon of the day after they are received.

New Customer

All new orders shall be processed and entered on file for shipment within one workday.

3. COST

Investment to Equipment Cost

Average inventory investment at cost shall not exceed 5 percent of the cost of equipment supported.

Investment to Service Revenue

Average inventory investment at cost shall not exceed 13 percent of total service revenues.

Investment to Parts Revenue

Average inventory investment at cost shall not exceed 24 percent of revenues for service parts.

Turnover

Average turns of inventory based on units shall be at least 3.5 per year. This may vary from at least 12 per year for fast-moving consumable supplies to less than one per year for insurance-class parts, with consideration for life-cycle stages.

4. STATUS

The status and expected or actual shipping date of every order received at supplier's Order Entry function shall be available through Order Entry within three minutes.

5. CUSTOMER EASE

Any customer contacting a parts stock facility or supplier Order Entry function shall receive assistance necessary to accurately order any replacement part with 95 percent success rate on the first request and 100 percent by the second selection.

6. REPAIRABLES

There shall be a maximum of 30 calendar days from removal through return to good stock.

7. SCRAP

The amount of parts scrap per year shall be no more than 8 percent of the value of new service parts added that year, or 3 percent of the value of all service parts inventories.

8. COMPLAINTS

Problems related to parts that cause customers to contact management shall be fewer than 0.1 percent (one complaint per 1000 orders).

3

Unique Differences in Service Parts

HE OBJECTIVE OF THIS chapter is to point out the characteristics of service parts support activities and to suggest improvements that should be made for better support in the future.

There are major differences in the way parts have normally been handled for manufacturing/production organizations and the way they need to be managed for service businesses. A lot of work has been done on production/manufacturing-type inventory, but relatively little has been carried out concerning the challenges of service parts. One of the things that service managers really need to do is to take a look at similar problems within service businesses.

The major service parts inventory factors are shown in Table 3-1.

Table 3-1
Service Parts Inventory Factors

1. Product life phases	7. Risk
2. Future demand variability	8. Essentiality
3. Causes for replacement	9. Flexibility
4. Repairables	10. Gain or loss
5. Multiple locations	11. Systems approach
6. People influence	

PRODUCT LIFE PHASES

The first main difference between the way parts should be analyzed for service and for manufacturing is the life phases through which service parts must go. There are five distinct phases of life. The first is *preproduction* for tests and training, with new equipment, very little historical data, some correlation with prior products, possibly laboratory tests, and reliability handbook predictions, but that is about all. Initial support is based on people's best engineering experience and educated judgment. Service parts, in spite of all our sophistication and technology, is still very much a people business. Experienced, educated, and dedicated people are vital to help manage service parts.

In the second phase, *product introduction,* a fundamental concern is having a basic supply of service parts. Provisioning must be carefully done using every possible source of information, and manufacturing must be pushed to deliver the required parts on schedule.

The third phase a product goes through is the ongoing *normal life phase,* which should be many, many years. History is recorded, computer models start to operate well, forecasts can be adjusted on the basis of good field predictions and what engineers say is going to happen to their design changes, and perhaps we will make a lot of money. Then, one day, production stops and the source of parts must change in the *post-production phase.* Eventually end of life comes in the *termination phase,* when the product has to be withdrawn from the marketplace. Very few people like to face the fact, early in a product's life, that their product may die. However, we must think about the idea because we must come up with very specific decisions for the end of life. Too many of us vacillate and say, "We're going to support it for ten years after the last one is built," or "We will support it for seven years after it's delivered to you." We all know products that are still in the marketplace 40 years later, and no one will say, "Sales, it's time you tell the customer that we are not going to support that model 44 any longer." It's an opportunity for Service to do a sales job on Sales and say, "It's your chance to get our latest, greatest product out there and replace that old one that's just limping along." Also, some service organizations evaluate parts on two- to five-year schedules during later years to plan discontinuance. So there are life phases in product support that do not generally concern manufacturing.

Future Demand Variability

Future demand variability is not like scheduling production. Manufacturing demands for parts tend to be deterministic given a bill of materials and a build schedule. Service parts demands are probabilistic. Service is a business that is going to flex. We make educated judgments, but we are predicting and forecasting the future, so we're working way out there with some additional complications. We must try at least to forecast what future demand is going to be. Some computer programs will do that, but a computer is only as good as history interpreted into the future. The computer cannot accurately predict what's going to happen unless we provide it with information that is modeled to show how the real world is going to operate. This is one area in which a well-educated and financially experienced person should be managing parts. A good person can handle up to 12,000 line items, although 3000–8000 is typical. As few as 300 line items may be a full load if these are repairables with a high activity rate and account for a large percentage of the business.

We must have a working relationship with Engineering and Marketing in order to know in advance what changes are coming. A lot of information is informal; you cannot force such communications into an ultraformal relationship. Service management should insist on sign-off of engineering change orders (ECOs). Any field engineering organization that does not have sign-off of ECOs is never going to be operationally effective. The issue must be forced, but there are many issues and communications that are much better handled over a cup of coffee... "Well, Sam Design Engineer, what changes are you planning? What's going to happen beyond the formal paperwork? Marketing, when is that product going to be terminated? Do you expect the superfuser really to install 3000 units? Are we planning a big promotion that will need a lot of consumables, or are we getting into a peculiar change in the market, or a new use for the product that is going to alter the usage of those parts? Are products suddenly going to be stocked at different locations, or shipped to remote locations whereas equipment has been geographically centered up to that point?" These things must be known and only a human can tell the computer what's going to happen. Otherwise it's the GIGO problem, which now says "garbage in, gospel out" because it's coming from a computer. Data are not necessarily accurate just because the data are in printout form. If you lie to a computer, it will lie to you. People have to help determine causes, risks, and ranges of future demand variability.

Causes for Replacement

The causes for replacement of service parts are not all attributable to failure. Many parts are replaced because diagnostics are inadequate, the manuals were incorrect, or the tech rep couln't find out what the problem was and so pulled out the easiest thing to get at. The cause was really a loose connection on the board, but when the tech rep replaced the whole board, the problem went away so the decision was that it was a defective board. The U.S. Air Force and most civilian industrial organizations find that 30–45 percent of all their electronic modules coming back as failed are actually good. A lot can be done to improve design of equipment and the support hardware and software, and to train maintenance personnel in how to troubleshoot problems.

There are many reasons why parts are replaced other than for reasons of failure. Training causes many failures when short circuits or overstress conditions occur. Inept operators damage equipment. Even willful damage sometimes takes place. And occasionally parts are replaced twice owing to incorrect installation on the first repair.

Testability is one of the technologies that must be improved in the near future in order to accurately test and diagnose the hardware and the software and tell exactly what is wrong. Eventually simple diagnostics should tell the operator go/no go, because more maintenance must be done by the operator. There will not be enough good technicians available to send out on every service call on every product in the world of the future.

Travel is both expensive and wasteful. Failure causes should be maintainable so they can be easily corrected by an operator at the location.

MULTIPLE LOCATIONS

Figure 3-1 shows the complexity of a typical service parts network. Most field service representatives have parts in their cars. Parts are also stored at district, area, branch, zone, or

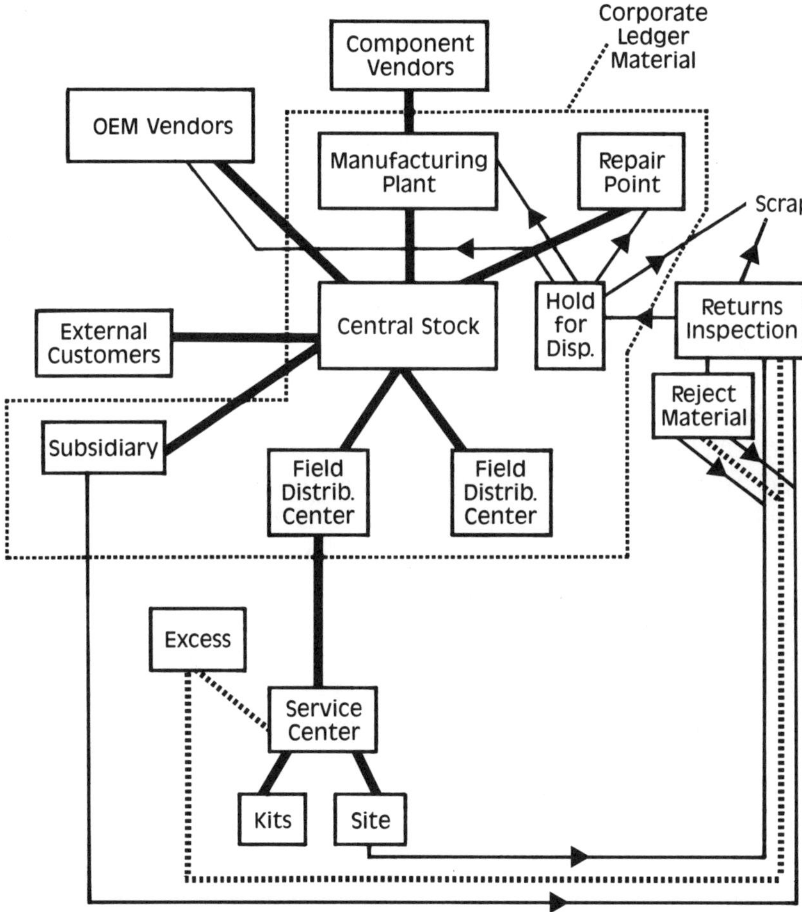

Figure 3-1
Material Flow—Service Parts

regional distribution centers—whatever the facilities are called. Service parts is a much more complex problem than having parts in one or two manufacturing locations. There are advantages to this if management can learn to utilize those multiple locations, but more locations usually mean increased parts quantities and costs.

There are economies of scale. If all parts are put in a centralized location where they can all be seen in one big shelf area, fine. When parts are distributed to the four winds, visual control is gone. Probably fewer than 5 percent of all service organizations have accurate knowledge of what is in car stocks. However, with the advent of distributed processing and computer systems, perpetual inventories can be kept. It is vital to control inventory by quantity. Parts cannot be controlled well by a once-a-year physical inventory that may not be done correctly anyway—at the holiday season, in lackadaisical fashion, over the weekend when people would rather be elsewhere. Once-a-year physical inventory often causes morale, accuracy, and reliability problems. As computer power becomes less expensive and more sophisticated, the link between maintenance reports and the parts used helps to improve effective control of inventory.

Management must keep current on what parts are where. To do that, there are two initial computer reports that will really help. One of these, the "New Item Candidate" report, says, "These are the parts that we use, but we have not stocked." If parts are requested, they should be on hand. If an emergency order was required, why weren't the parts in stock? There we have the information that says, "We should be carrying those parts."

On the flip side of that situation are the parts that we have in stock but did not use in the last six months, for example. We didn't use them, so why are we carrying them around in every car trunk? Turnover and use reports provide the information that enables us to go back and take a good look at both supply and demand. The holding of parts and the actual use of those parts has to be tied together.

PEOPLE

The influence of people is major in our service system. A production-line approach to service helps to standardize a lot of things in the service area and give people direction. If people don't have direction, they'll do their own thing. Intelligent service managers need to determine the one best way to do things, gain the participation of employees, and work with them so they will do their jobs in the best possible fashion. Service parts is a people business.

People need to be trained. All parts people should be educated about the service parts business, and about mathematics and how to use the basic economics and statistics that are necessary to understand supply and demand. This knowledge should be taught at a fundamental level so that all service representatives can learn to manage parts for their particular territories. An observer on a recent field improvement program watched a tech rep diagnose a defective microswitch, take the switch apart, and spend 1½ hours rebuilding the switch, including soldering internal components. The rep had four calls waiting, and had three microswitches in the car trunk right outside the door. He spent 1½ hours because his manager had recently told him that his parts were costing too much. If there hadn't been a well-qualified observer watching, it would never have been believed! That's a case of pure stupidity, or mismanagement—there can be many adverse words connected with it. The tech rep said he was doing what his manager wanted him to do. He truly thought he was doing the right thing, and yet that equipment probably failed the next day because he had reduced the reliability of the switch. And he wasted 1½ hours that could have been billed at the rate of about $60 an hour. By the time that second call was completed, he probably had invested $150 in a $3 switch. And yet might not your reps do the same thing? Do your people have the right kind of guidance to suggest why that was a bad thing to do? That tech rep worked for the service division of a Fortune 500 company!

PARTS DEMAND–SUPPLY CYCLE

Anytime a tech rep comes into the office in the morning to pick up parts, at least 45 minutes of productive time will be lost, and probably more. Not only are tech reps not starting their first call of the day at starting time, but they're often wasting time talking with other employees, having a cup of coffee, and then starting to drive to their first customer. A lot of time is lost. Get those parts into the tech rep's hands with a "milk run" that travels a regular route, or by direct shipment, or in whatever way gets the parts efficiently to the point of need.

It is tremendously important that field engineering, field service planning, service program management—people with field experience—get in on the early design stages of any product. It is not possible for the typical design engineer to understand all the environments that products are going to encounter in the field. It is not possible to learn experience from a textbook. You must get out in the field, you must see it. Take engineers by the hand to user locations and show them how the products and parts are being used in the field. Show them what the customers do with parts. Listen to customers. Watch casual operators try to manipulate equipment. It's a challenge because engineers are always working on high priorities, albeit perhaps not effectively. They may be working on the wrong thing. Service should provide guidance and arrange for experienced service engineers to work closely with product engineers during new-product design and development.

Another alternative is to present a "show-and-tell" in-house. Some very effective things can be done with photographs, 35-mm slides, and audio-visuals during a lunch hour. Have a lunch-time presentation in the local cafeteria. Show some of the situations in which the equipment is used, how parts are used, and how they're handled. Show all the steps of the demand-and-supply cycle that the parts go through so people can pre-act instead of just react.

RISK

Manufacturing must have every part on the assembly bill of materials. No service organization can afford to stock 100 percent of possible needs. Risk is a fact of life with service parts. Service managers have to learn to deal with risk. Judgment is rarely going to be right all of the time. An organization that carries all the parts ever needed must be richer than Rockefeller ever was. The typical service manager can become very emotional over that customer who cannot be supported because of the lack of a part. Ninety-five percent availability of parts means 5 percent are not going to be immediately on hand. Risk can be measured as the percent of orders not filled within the specified time limit. For example, if 98 percent of field replaceable units (FRUs) are to be available within 24 hours, then the risk is $100 - 98 = 2$ percent. Learn to live with that kind of quantification and risk in order to manage the parts business.

ESSENTIALITY

Essentiality attempts to concentrate attention on the most critically important parts. It is often a four-point system. Basically, the class number 1 is assigned if the concern is safety or legality, or if the part is likely to require complete shutdown if it fails. A safety valve is obviously a 1 item. A number 2 rating means the equipment is not operational. A number 3 means slightly degraded, substandard, slow speed, low quality. A number 4 is a minor cosmetic item. Class 1 and 2 parts will be car stock if they are frequently used and are cost effective. Rarely should a tech rep ever carry a 4 rated part. The next service call is usually adequate to replace that glass, or panel, or to take the spray paint and touch it up. Note that essentiality alone does not answer the question, "Stock or not?" Cost and frequency of need must also be considered.

The author first described essentiality in a 1971 paper on planning spare parts for preproduction products, which is partially covered in Chapter 8. However, only 5–10 percent of service parts systems today gain the advantage of essentiality. Assign an essentiality rating to every part. A part should be given an essentiality rating when it is entered into the information system by a service engineer who understands how that part is being used in the machine.

The machine itself also has an essentiality rating so we know how important it is. If it's a life-support machine, it gets a very essential 1 rating. If it's equipment such as an overhead projector, it's less critical. If the projector light goes out, it's out of commission but the failure certainly isn't going to be cata-strophic. That light would be an essentiality 2 part because if the light burns out, the projector can't be used. Fortunately, companies such as 3M provide two light bulbs so that one is in use and the other is redundant, but can be quickly switched into use if the first fails. Maintainability is helping where reliability isn't so great, because a typical projector bulb only lasts about 40 hours. Essentiality then also can be used at a working operational level. A particular piece of equipment may be essential to a specific customer. If a medical center has four sterilizers, the situation is not critical if one of them fails. If there is only one in a small rural hospital, that one is more essential to the small hospital than any one of the four is to the medical center.

Essentiality considerations are important to service, but they are not so useful to manufacturing. To build a machine requires all parts, with little recourse. Everybody in service should be using essentiality. Every part and every piece of equipment supported should be rated according to its essentiality. Essentiality is a major factor that is not yet universally considered in service, but should be.

FLEXIBILITY

The ninth factor is flexibility. There are many opportunities in service to use management information and to make use of the multiple part locations to find the needed parts. Go to the next higher assembly. Look to other products on which the part is used. Repair a damaged part. Even, as a last resort, canibal-ize (dismantle a piece of equipment for its parts). There are many different places where a part can be found, if the infor-mation is available on where to find it. Management information systems in the future should disclose perpetual inventory to everyone involved, right down to the field representa-tive/dealer level so that it is known on at least a weekly basis, and preferably daily, where those parts are—from regional distribution centers and district stocks to the individual person in the field.

The challenges and opportunities are ripe for flexibility, but service operations often build on quicksand in that area. If we say, "Here's your recommended stock list, but go ahead and stock whatever you want," there is too much variability in that approach. Too often stocks are limited as a result of artificial measures rather than accurate business-level forecasts based on sales, turns, and benefits of the recommended stock list.

Several large companies give their service reps and dealers a recommended stock list and require stocking within 10 percent of those parts. The stocking lists are based on the product models in the territory, and the population and use of the equipment being supported. Push-versus-pull inventory is one of the techniques that goes straight to the idea of saying, "Management knows better what inventory you ought to carry." That narrow position can cause problems if a tech rep suddenly has to go on a call out of the territory and may not have the parts for the equipment to be serviced. "Based on good historical information and accurate projection of the future, these are the parts and the quantities you should carry." Stocking recommendations for local inventories and car stocks can be greatly aided by the use of worldwide demand data.

The reps must have confidence that they can get resupply on those parts. One of the biggest single factors that causes tech reps to stock excess parts is lack of confidence in being resupplied. It is much more important to have a consistent supply system than one that sometimes is very fast and other times is very slow. It does very little good to have one foot in the oven and the other in a block of ice. On the average you may be statistically comfortable, but you're burning up one foot and freezing the other! A lot of our parts support is done that way. Somebody yells "Wolf!" and all of a sudden the parts are supplied on emergency order overnight, but then normal orders take several weeks. Trust will be lost and you can't blame people for wanting to caché away a supply of parts if they don't have confidence in the system. Consistent, good support is a challenge, and a tremendous opportunity for management. Once confidence improves, stock that was once thought essential will be given up as excess.

GAIN OR LOSS

The tenth factor is the gain or loss that each of those parts will bring. Most organizations supply parts at a profit. An ABC inventory level that strictly multiplies the cost of the part times the number used is not adequate. Would you rather have one $1000 part or a thousand $1 parts? Most service reps probably will opt for the thousand $1 parts because the chances of their getting what they need are much higher than if you have one $1000 circuit board. Further, both absolute and relative profit should be considered. For example, a $10 part in inventory might be marked up seven times and sold for $70. There's a difference of $60 that is profit. Wouldn't it be better to have that high-markup part available to sell to somebody versus a part that sells for double the price with gross profit of just $10? Would you rather make $10 or $60? To most persons, $60 is the answer, if the probability of needing it is about the same. So there is a positive, optimistic profit-oriented gain.

On the other side—the loss side, and preventing the loss—look at how much time reps in the field take to get parts. Many persons say, "That's a necessary evil." If fixed-price contracts or leases require parts, then any money saved is desirable. There is very accurate information from service improvement programs that proves that parts are costing a lot of money. Some accurate parts-related time data from a well-run large service organization are shown in Table 3-2. These data are based on over 700 hours of observation of 65 tech reps.

Table 3-2
Tech Rep Parts Time

	Average Minutes	Number of Occurrences
Obtain part number	6	22
Requisitions	8	67
Travel to get part	15	23
Checking on orders	13	18
Return to stock or defective	18	14
Cannibalize	15	4
Pricing	8	2
Total: 4.0 percent of paid time		

If a part is needed and it's essential but not on hand, the service rep has to contact the branch and find out where it is. Perhaps another rep has one. At the very minimum, the rep is going to have to get the part number, to prepare a requisition, to travel to get the parts, and to go back to the customer. This will probably take at least an hour. What's that hour worth to you? Service labor costs \$45–\$100+ an hour these days. Multiply your rate by time, plus the cost of transportation, and that quantifies a large loss that can be avoided. Parts can be traded against people. If service reps always have the right part, it may cost more, but it's going to make the reps much more productive and it's going to make customers a lot happier—and you're going to do more business and make more money.

Probably 2–3 percent of a tech rep's time is normal to devote to working with parts, getting part numbers, obtaining them, and so on. Most organizations are probably in the 5–8 percent area of time spent to get parts. Take that difference and multiply it by an eight-hour day and by the labor rate charge, and it's going to be probably at least \$10 a day in value chasing parts that could better be revenue-producing time! A lot of parts-related times are hidden today in organizations. A lot of effort and time—the stockkeeper calling back and forth, the dispatcher chasing parts, managers arranging transfers—is used in handling parts, and most people don't realize it. That's another opportunity for improvement.

SYSTEMS APPROACH

Finally, a systems approach is vital. Put all the elements together. Look at the total life cycle. Plan products from the very beginning and design the support system with the product rather than afterwards having to react and support the design. Use service-level data to measure performance and adjust the stock levels to meet management's goals. As an example, the service level of a multiechelon distribution system can be 99 percent even though the fill rate at each location is 70 percent or less, as shown in Table 3-3.

Table 3-3
Multiechelon System Performance

Stock Level	Number of Orders	Number Filled	Percent Filled	Cumulative Percentage
Local	100	70	70	70
Regional center	30	20	67	90
Central stock	10	7	70	97
Manufacturer	3	2	67	99
(Not filled on time, back-ordered)	1	1	100	100

Establish a total approach and an integrated service parts management system!

4

RECORD SYSTEMS

UNIQUE IDENTIFICATION

A FUNDAMENTAL START for managing service parts is to assign each stockkeeping unit a unique part number. The part number should be all numeric—contain no alpha characters, dashes, or spaces; should not attempt to identify the group or class of the part; and should be as short as possible. Why? Simplicity enhances accuracy and speed. The part number is the key to all record keeping. Most data entries are made by people who either write down part numbers or push VDT (video display tube) keys. The fewer digits there are to be remembered, the better. Error rate increases as the number of digits increases. Most systems can operate with five digits, allowing up to 99,999 line items. More items than that, unless supporting a very large organization, indicates a problem with items. Data entry can be done at high speed using the numeric keypad on computer terminals if all the entries are numeric. Injecting a space or alpha character requires hand movement to that key and disrupts the process. The minus (–) key on most numeric pads can substitute for the dash, but should be necessary only in long numbering systems. Alphanumeric (*X*-field) computer programs require all five digits to be entered, even if they are 00001, so that the part numbers will sort in the proper order. An all-numeric (9-field) program will right-justify entries and save effort by requiring entry of just 1 instead of 00001. If a small organization can foresee a need in the next five years for fewer than 9999 part numbers, then it should use only a four-digit system. Suppress leading zeros whenever possible, since the human eye takes time to distinguish a zero from other characters, but does not need the same time to recognize spaces.

COMMODITY CODES

A question often raised is, "Should not the part number identify the kind of equipment it is?" The federal stock-number system does that, and a very small system running purely on manual cards could also. In between these very small and very large systems, the costs are generally much greater than the benefits. The typical method of identifying a commodity is by group and class. Table 4–1 shows some examples from the *Federal Supply Classification Cataloging Handbook.*

The advantage of having a group and class embedded in the part number is that the stockkeeping personnel can quickly learn to double-check the parts they are handling. If they are pulling part 61051234, for example, it should be a motor. If it is a relay, they should easily reject it as a mistake. Another plus is the ease of dividing parts into catagories for purchasing and pricing attention. The overwhelming disadvantages include:

1. Extra digits are attached to a part number that must be used in every transaction. The few extra seconds necessary to speak or to write down the number and then to check those digits can add up quickly.

2. The commodity group class may change or the part may fit into several commodities. Items common to multiple products should have a common part number. It would be foolish to have separate stocks of the part with only the part numbers different.

3. If different part-numbering systems ever have to be in-

tegrated, the commodity designation will probably cause trouble. Few people have the foresight to know what future products may be developed, and also what numbers to assign to which categories to avoid running out.

A combination that can be advantageous for organizations with high outside customer service components is to define the group and class for each part and append them as prefixes to the part number. Thus if customers are ordering a new wick for a kerosene heater, which will be a frequently ordered item, the added digits can help assure that the customer has ordered and will receive the desired part. The part number should not require the group/class prefix in order to be unique. In other words, the same part number should never be duplicated in an-

Table 4-1
Examples of Commodity Codes

Group		Class	
25	Vehicular equipment components	2510	Cab body and frame structural components
		2520	Power transmission
		2530	Brake, steering, axle, wheel, and track
		2540	Furniture and accessories
		2590	Vehicle miscellaneous
28	Engines, turbines, and components	2805	Gasoline, reciprocating
		2815	Diesel
31	Bearings	3110	Bearings, antifriction unmounted
		3120	Bearings, plain unmounted
		3130	Bearings, mounted
55	Construction materials	5510	Lumber and basic wood
		5520	Millwork
		5530	Plywood and veneer
		5630	Pipe and conduit, nonmetalic
61	Electrical components	6105	Motors, electric
		6110	Electric control equipment
		6115	Generators
		6120	Transformers
		6135	Batteries

other group/class. The commodity group and class may be kept, of course, in the part record as separate fields, which will allow a sort on those parameters for purchasing and stocking use. The group and class are related to the noun nomenclature, which is a preferred way of operating. A typical list of parts by commodity code is given in Figure 4-1.

NOUN NOMENCLATURE

Noun nomenclature means that every part should fit into one of about 550 noun categories. Table 4-2 lists the nouns used by several public utilities, metal and chemical refineries, manufacturers, and service organizations.

By "pigeonholing" every part into one, and only one, of these descriptions, it is possible to find the desired item by using the common word by which it is normally described. A ten-digit alpha field is enough for most descriptive nouns. If the word happens to be longer, such as "transformer," simply use the first ten letters and enter it as "transforme." About 95 percent of all the nouns fit into ten digits or less, and the rest may be easily truncated, so a field of ten is a good choice. The value of noun nomenclature was demonstrated when the author found the suits worn by maintenance personnel in different locations of a stockroom labeled variously: clothing, coveralls, pants, trousers, and overalls.

Following the noun should be descriptive adjectives that define the part in detail. A 40-character free-form field is recommended, and some people will want more. Use common abbreviations where possible. The objective of the noun and description is to show on the parts card or computer printout enough information so that someone needing the part can quickly determine whether that part is already in stock. Too often, review is cumbersome and parts are ordered even though they may be available internally. The sequence of descriptive words should be: function, type, material, color, and size.

```
          A L L     P A R T S     F O R     A    C O M M O D I T Y

ENTER COMMODITY CODE:   3110  DESC: BEARINGS, ANTIFRICTION, UNMOUNTED
ENTER SCREEN VIEW (S) / PRINT (P): S

STOCK NUMBER     NOUN        DESCRIPTION

        3002    BEARING      ,ROLLER,BALL,STNLS,3.6MM
        3004                 ,ROLLER,BALL,STNLS,3"ID,5"OD
       24355                 ,ROLLER,STNLS,W/SEAL,3"
       99652                 ,ROLLER,PHOSBRONZ,1/2"
      123456                 ,ROLLER,PHOSBRONZ,3/4"
      124434                 ,BALL,TIMPKIN MOD23,3.5MM
      124435                 ,BALLS FOR SN 124434
      124436                 ,RACE FOR SN 124434
```

Figure 4-1   Typical List of Parts by Commodity Code

Table 4-2
Part Noun Descriptions

ABRASIVE	BOARD	CEMENT
ABSORBER	BOLT	CHAIN
ACCUMULATO	BOOK	CHANNEL
ACTUATOR	BOOM	CHART
ADAPTER	BOOT	CHASSIS
ADHESIVE	BOTTLE	CHEMICAL
ADJUSTER	BOWL	CHISEL
ALARM	BRACE	CHOCK
AMPLIFIER	BRACKET	CHOKE
ANALYZER	BRAID	CHUCK
ANCHOR	BRAKE	CHUTE
ANODE	BREAKER	CIRCUIT
ANTENNA	BRICK	CLAPPER
ANTIFREEZE	BRIDGE	CLEANER
ANVIL	BROOM	CLEAT
APERATURE	BRUSH	CLEVIS
APPLICATOR	BUCKET	CLOTH
APRON	BUFFER	CLOTHING
ARBOR	BULB	CLUTCH
ARM	BUMPER	COATING
ARMATURE	BURNER	COIL
ARMOR	BURNISHER	COLLAR
ATTENUATOR	BURR	COLLECTOR
AWL	BUSHING	COMPRESSOR
AXE	BUTTON	CONDENSER
AXLE		CONDUCTOR
	CABINET	CONDUIT
BACKING	CABLE	CONE
BAFFLE	CAGE	CONNECTOR
BAG	CALIPER	CONTACT
BALANCER	CAM	CONTAINER
BALL	CAMERA	CONTROL
BALLAST	CAMSHAFT	CONVERTER
BAND	CAN	CONVEYOR
BAR	CANOPY	COOLANT
BARREL	CAP	COOLER
BEARING	CAPACITOR	CORD
BED	CARD	CORE
BELLOWS	CARRIAGE	CORNER
BELT	CART	COUNTER
BINDER	CARTRIDGE	COUPLING
BIT	CASE	COVER
BLADE	CASTER	CRANK
BLANK	CASTING	CRIMPER
BLANKET	CATCH	CROSSARM
BLOCK	CATHODE	CRUCIBLE
BLOWER	CAULK	CRYSTAL
BLUEING	CELL	CURTAIN

Table 4-2
Part Noun Descriptions (Continued)

CUSHION
CUTTER
CYLINDER

DAMPER
DECODER
DEFLECTOR
DEGREASER
DETERGENT
DEVELOPER
DIAL
DIAPHRAGM
DIE
DIFFUSER
DIODE
DISCONNECT
DISINFECTA
DISK
DISPENSER
DISPERSANT
DISPLAY
DISTRIBUTO
DIVIDER
DOLLY
DOOR
DRAIN
DRESSING
DRILL
DRIP
DRIVER
DRUM
DRYER
DUCT

E-RING
EARPHONE
ECCENTRIC
EJECTOR
ELBOW
ELECTRODE
ELECTROLYT
ELEMENT
EMMITER
ENCODER
END
ENVELOPE
EPOXY
ESCAPEMENT

EXCHANGER
EXHAUST
EXPANDER
EXTENSION
EXTINGUISH
EXTRACTOR
EYE

FACE
FACING
FAN
FASTENER
FEED
FERRULE
FIBER
FIELD
FILAMENT
FILE
FILLER
FILM
FILTER
FINDER
FINGER
FITTING
FIXTURE
FLAG
FLANGE
FLAP
FLARE
FLASHBAR
FLASHER
FLOAT
FLUID
FLUX
FOAM
FOLLOWER
FOOT
FORCEP
FORK
FORM
FRAME
FUEL
FUNNEL
FUSE

GAGE
GAS
GASKET

GATE
GEAR
GEARBOX
GENERATOR
GLAND
GLASS
GLASSES
GLOBE
GOVERNOR
GRATING
GREASE
GRID
GRILL
GRIPPER
GRIT
GROMMET
GUIDE
GUN

HAMMER
HANDLE
HANDSET
HANDWHEEL
HANGER
HASP
HEADSET
HEATER
HEATSINK
HELICOIL
HINGE
HOE
HOIST
HOOK
HOSE
HUB

IDLER
IGNITOR
IMAGER
IMPELLER
INDEX
INDICATOR
INDUCER
INDUCTOR
INJECTOR
INK
INSERT
INSERTER

Table 4-2
Part Noun Descriptions (Continued)

INSTRUMENT	MIXER	PLUNGER
INSULATOR	MONITOR	POINTER
INTERFACE	MOP	POLE
INTERPOSER	MOTOR	POLISH
INVERTER	MOULDING	POSITIONER
	MOUNT	POST
JAW	MUFFLER	POWERSUPPL
JET		PRINTBAND
JEWEL	NAIL	PRINTER
JOINT	NAMEPLATE	PROBE
	NAPKIN	PROCESSOR
KEY	NECK	PROPELLER
KNIFE	NEEDLE	PULLER
KNOB	NIPPLE	PULLEY
	NOZZLE	PUMP
LABEL	NUT	PUNCH
LADDER		PURIFIER
LEAD	O-RING	
LEG	OIL	RACE
LENS	OILCAN	RACK
LEVEL	OILER	RADIATOR
LEVER	OINTMENT	RAIL
LIFTER	OSCILLATOR	RAKE
LIGHT	OUTLET	RAM
LIMITER		RAMP
LINE	PACKING	RATCHET
LINER	PAD	REACTOR
LINK	PAINT	READER
LOCATOR	PANEL	RECEIVER
LOCK	PAPER	RECEPTICLE
LOCKPLATE	PAWL	RECORDER
LUBRICANT	PEDAL	RECTIFIER
LUBRICATOR	PEN	REDUCER
LUMBER	PENCIL	REEL
	PENETRANT	REFLECTOR
MAGNET	PHOTOCELL	REGULATOR
MANDREL	PILOT	RELAY
MANIFOLD	PIN	RELEASE
MANTEL	PINION	REMOVER
MARKER	PIPE	RESIN
MASK	PISTON	RESISTOR
MAT	PIVOT	RETAINER
MESH	PLATE	RHEOSTAT
METAL	PLATFORM	RIBBON
METER	PLENUM	RIM
MICROPHONE	PLIER	RING
MILL	PLOW	RISER
MIRROR	PLUG	RIVET

**Table 4-2
Part Noun Descriptions (Continued)**

ROD	SPRING	TURNBUCKLE
ROLLER	SPRINKLER	
ROPE	SPROCKET	VALIT
ROTOR	SPUD	VALVE
RUBBER	STAKE	VANE
	STANCHION	VISE
SADDLE	STAPLE	
SAMPLER	STARTER	WASHER
SAW	STATIC BAR	WAX
SCALE	STEP	WEIGHT
SCALPEL	STOP	WHEEL
SCANNER	STRIKE	WIDGET
SCOPE	STUD	WINCH
SCRAPER	STYLUS	WINDOW
SCREEN	SUMP	WIPER
SCREW	SUPRESSOR	WRENCH
SEAL	SUSPENSION	WRINGER
SEAT	SWITCH	
SEGMENT	SWIVEL	YOKE
SELECTOR		
SENSOR	TABLE	
SEPARATOR	TAP	
SET	TAPE	
SHAFT	TAPPET	
SHEATH	TELEPHONE	
SHEETING	TELESCOPE	
SHELF	TENSIONER	
SHIM	TERMINAL	
SHOE	TESTER	
SHUNT	THERMOCOUP	
SHUTTER	THIMBLE	
SIGHT	THYRISTOR	
SIGN	TIE	
SLEEVE	TIMER	
SLIDE	TONER	
SLING	TOOL	
SNUBBER	TOOTH	
SOCKET	TORCH	
SOLDER	TOWEL	
SOLENOID	TRACK	
SPACER	TRANSDUCER	
SPEAKER	TRANSISTOR	
SPINDLE	TRANSMITTE	
SPLASHPLAT	TRAP	
SPLINT	TRIP	
SPONGE	TRIPOD	
SPOOL	TRUNNION	
SPOON	TUBE	

NATIONAL STOCK NUMBER

The National Stock Number (NSN) system is one of the most comprehensive in existence. It is well illustrated in *Military Handbook 5* with supplements. The U.S. government thinks it important to know where a part is manufactured as well as the group and class that defines what kind of material it is, and a unique identifier. The North Atlantic Treaty Organization (NATO) modified the system in order to make it acceptable around the world. Then, expecially for commercial purposes, the Australian government developed Auslang, which is a further extension of the system. This is especially necessary for Australia because it relies on many other countries for manufactured goods, especially to support the country's mining and metal-refining operations that are often several days removed from the closest source of resupply. These series of manuals are all cross-referenced so that commodity code groups and classes can be found from the noun and description, and vice versa. For commercial and industrial purposes, many of the groups and classes for weapons systems should be deleted and a few additional ones are required. The Du Pont Corporation found that its effort to develop a commodity code group/class and noun nomenclature system necessitated the labor of five people over a year, and the system underwent further refinement during actual use.

FORM, FIT, FUNCTION

These three parameters are used to decide whether a part is different from others. Consider a printed circuit board (PCB) that the engineers want to call "part 456, revision B." This PCB looks the same as its previous cousins, fits the same card cage and connectors, but has the additional capability of automatically sensing and accommodating varied transmission rates. The question is whether it should be called revision B or should receive a new part number. This is a boarderline situation that is best resolved by using a new part number. It will provide better control, unless the organization tracks parts by unique revision levels. If the latter is done, the revision is treated much like extra digits on a part number. Steiger Tractor, for example, uses a T1 or T91 suffix to indicate advancement of a part. Consideration should be given to the fact that as soon as technicians discover that one variation of a part is more reliable or functions better than another, they will push for the improved version. Inventory managers should assure that stocks of the older part are used whenever possible before the new part numbers are released.

BILL OF MATERIALS

All manufacturing operations start with a bill of materials (BOM). This list of components in equipment is "exploded," starting with the top system level, into subsystems, then assemblies, then subassemblies, and finally piece parts. Every component then, except the top level, has a "next higher assembly" and also can be categorized "where used." Factory maintenance and field service have little use for the complete manufacturing bill of materials. However, they do need to know the maintenance replaceable parts. These are often referred to as line replaceable units (LRUs) or field replaceable units (FRUs). Individual electronic components soldered onto a PCB are not normally field replaceable. Complete PCBs, exterior panels that are subject to damage, and power supplies would be listed as field replaceable parts. On computers and office equipment, the steel frames and welded assemblies are not normally field replaceable, but agricultural and industrial operations frequently replace such parts.

The decision to catalog a part number as available for service is different from the decision to hold service stock on that part number. Exterior panels, for instance, are an essentiality 4 item and would not be stocked in the field service supply system. Should a need arise for the part in the field, an order would be placed and would flow normally through the order system. When the order reached the analyst, a purchase order would be sent to manufacturing. For in-production items, the order can be filled quickly. Out-of-production items may require full manufacturing lead time.

In many service organizations, a spare parts list (SPL) is developed in the provisioning process and maintained on a database for access throughout the life of the product. When the last product is removed from field service, the part numbers associated with that product should be purged from the inventory.

Reference designators are also useful physically and logically to locate parts within equipments, or even equipments within systems. For example, all resistors may be coded with an R and then sequentially numbered as R1, R2, ... Equipments within a large facility might have a system component designator such as "12 MOV 1234," meaning that this is in the System 12 Motor Operated Valve 1234. That is both a functional and a physical location and may be cross-referenced by the "where used" list. A major value of "all parts used on an equipment" is to display any parts stocked. It also identifies which parts should be disposed of along with equipment that is being removed. Too often, equipment is scrapped and sold but the support parts remain on the stockroom shelves until someone notices the thick coat of dust and lack of activity. Effective asset management will offer the support parts for sale or scrap together with the capital equipment. The "where used" information helps assure that parts are prioritized for essentiality in relation to the equipment they support. Also, standardization is promoted by using the same part on several products. And if one of those products is being terminated, we want to assure that the parts are still available to support other equipment on which they are used.

SPECIAL CONTROLS

The number of data elements for service parts tends to expand as the effort to control service parts increases. It is not unusual to have 50 or more data elements for each part number on the Parts Master Record. An example of a field list for a Parts Master Record is shown in Table 4-3. Normally these elements are entered through the cataloging organization. Data elements are entered in the part record in several ways:

1. By direct entry through cataloging organization
2. By data handoffs from suppliers
3. By data handoffs from the warehouse control system
4. By internal calculation

Table 4-3
Parts Master List

Field	1.	**Part Number**
List	2.	**Estimated Base Cost Code**
	3.	**Description**
	4.	**Unit of Measure**
	5.	**Base Cost**
	6.	**Retail Price Code**
	7.	**Retail Price**
	8.	**Retail Note**
	9.	**International Price**
	10.	**International Price Code**
	11.	**Packaging Code**
	12.	**Source Code**
	13.	**Sellback Code**
	14.	**Vendor Number**
	15.	**Vendor Part Number**
	16.	**Repairability Code**
	17.	**Percent Repair Cost**
	18.	**Product Code**
	19.	**Status Code**
	20.	**Comment Code #1**
	21.	**Comment Code #2**
	22.	**Provisioning Analyst Code**
	23.	**Provisioning Code**
	24.	**Provisioning Date**
	25.	**Criticality Code**
	26.	**Status**
	27.	**Equipment Application #1**
	28.	**Equipment Application #2**
	29.	**Equipment Application #3**
	30.	**Equipment Application #4**
	31.	**Equipment Application #5**
	32.	**Capital Equipment Code**
	33.	**Field Change Order (FCO)/Engineering Change**
	34.	**FCO Class**
	35.	**Tariff Code**
	36.	**Canada Tariff Code**
	37.	**International Repair Credit**
	38.	**Responsible Inventory Analyst**
	39.	**Vendor Type**
	40.	**Primary Repair Facility**
	41.	**Secondary Repair Facility**
	42.	**Field Repair Facility ID**
	43.	**Part Code**
	44.	**Upgradeable Code**

**Table 4-3
Parts Master List (Continued)**

Field		
List	45.	**Critical Code**
	46.	**Retail Print**
	47.	**Chain Phase Flag**
	48.	**Size**
	49.	**Hit Activity**
	50.	**Stock Keeping Unit**
	51.	**Pick Stock**
	52.	**Warehouse ID**
	53.	**Emergency Reserve Quantity**
	54.	**Repair Center Only Code**

In all cases data quality must be a primary concern. Data maintenance is an ongoing task for the cataloging function. Exception reports that contain data considered abnormal aid in maintaining data quality. In addition to the fields entered to the database from the cataloging organization, other data enter the Parts Master Record. Some fields are updated by supplier data:

—Purchase lead time
—Repair lead time
—Production status—in/out of current production

Some fields are filled by entry or change of related data:

—Standard cost date of last change
—Base cost date of last change
—Retail price date of last change
—International price date of last change

Some fields are calculated and displayed:

—ABC class code
—Maximum stock level
—Safety stock level
—Economic order quantity
—Economic repair quantity
—Reorder point quantity
—Use to date
—Returns from users to date
—Purchases to date

Change control should be regulated by a central engineering function, with sign-off by maintenance or field service. Changes to part numbers, supersessions, revisions, and use of discrepant materials should all be approved by the affected users before implementation. Too often an administrative-type person may judge that these have no influence and yet the stockroom may have large quantities of parts on the shelf or be in the midst of negotiations for purchase or repair. Certainly tools, test equipment, and documentation must also be considered before changes are approved.

Quality control of parts requires special attention. At nuclear power stations, for example, every part will be categorized as quality assurance or not, and will have subgroups within the qualification. For example, 1E means a safety-related part that must be qualified to perform in specal environments. Concern arises when the same part is used in several locations. A relay, for example, may be a category 2 in many systems and cost $5.50. If it is to be qualified 1E category, that same part could cost $250. If the cost difference is relatively small, all quantities of a part should be bought according to the highest qualification and then may be used wherever needed without any special control. If, however, there are large cost differences, the environmentally qualified parts should have a different part number and be treated as completely different parts, which they are because their function is different. It may be useful to retain the "interchangeable with" information on the part record for the lower qualified part since a more costly, more highly qualified part could be used in the lower situation for emergencies. The reverse is not acceptable.

Environmental controls themselves are a concern with parts such as gaskets, O-rings, paint, chemicals, and even bearings that may deteriorate with age or in the presence of certain chemical vapors or temperature extremes. Bearings and rotating equipment such as motors should be revolved every few months to assure they do not develop flat spots. Lot control presents some unique problems that are presently best handled manually. The relatively few parts that do require shelf-life control, rotation, or inspection controls can be flagged on the computer or manual record and listed so a human can perform the required physical action. Table 4-4 shows another detailed Master Parts Record. Note that in this case quantity and location are on separate records.

**Table 4-4
Maintenance Parts Database**

Item	Abbreviation	Field	Item	Abbreviation	Field
Part identification	Part ID	9 (6)	Selling price	Sell $	99,999.99
Description	Noun X (10)	X (40) (S)	Cost	Cost $	99,999.99
Commodity code	Comm code	9999 (X)	Margin $	Marg $	99,999.99
Revision	Rev	XXX ?	Margin %	Marg %	99
Essentiality	Essn	9	Carrying cost %	Carry %	99
Value class	Valu cl	X	Carrying cost $	Carry $	9999
Service only	Svc only	Y / N	Make / buy cost	Order $	999
Stock locations	Stk loc	X	Return credit	Rtn cred $	999
Stock level %	Tgt lvl	99	Repair cost	Rpr cost $	999.99
Use / yr	Use / yr	99999	Vendor code	Vendor code	X (6) (S)
Lead days	Lead dys	999	Vendor name	Vendor name	X (20) (S)
Reorder point	ROP	999	Vendor Part ID	Vendor PN	X (12) (S)
Minimum cost quantity	EOQ	99999	Equipment used on E / P	Eqpt use	X (12) (S)
Measure unit	Meas	XX	Higher assembly	Hghr assy	X (12) (S)
Quality class	Qa cl	XX	Interchangeable parts	Intrchgbl	X (12) (S)
Environmental	Envir	X	Superseded by	Suprsd by	X (12)
Shelf life	Shelf lf	Y / N	Purchase order ID	PO #	X (20)
Outage	Outage	Y / N	Quantity on hand	Qty-oh	9999
Location	Loc	X (10)	Quantity on order	Qty-OO	9999
Date last invent	Date last	99 / 99 / 99			

```
            E N T E R    N E W    P A R T

STOCK NUMBER: 012345    DESC: CONNECTOR , INSULATED, MULT TAP, COPPER, #4 TO 350
                                     SPECIAL PURPOSE PER DWG 32F-21
COMM CODE: 5935    SUPRSD BY:        *WIDCO OR KIDDEL ONLY
QA CL: 1M   ESSN: 2   ENVIR: Y   SHELF LF: N   OUTAGE: N   VALUE CL: C
STD PARA: QA21            QA33          PAS1


USE/YR:       30   UNIT: EA   LEAD DAYS:   60   AVG COST$:        647.75    REPAIR: N
CARRY%: 30   CARRY$     227   ORDER$: 102   DTO:     PO#:                  LN:

CAL ROP:      11   ADJ ROP:           S/P PROD SYSTEM COMPONENT      INTRCH
CAL EOQ:      15   ADJ EOQ:      18    S        01107 C122          012346
                                      P   MOT
VENDOR CODE: G111
VENDOR NAME: GENEROUS ELECTRIC
VENDOR PART: GE2543F45LM2312

VENDOR CODE: L355
VENDOR NAME: LINK FOUGHT
VENDOR PART: LF23554P
```

Figure 4-2   Detailed Part Record Screen

```
        P A R T    R E C O R D    A N D    A V A I L A B I L I T Y

PART ID  70001              DESC BEARING      ,ROLLER, STAINLESS, 7MM

ON HAND      5    LOCATION  A22B1       VENDOR      VENDOR       EQUIPMENT
ON ORDER     0    UNIT OF ISSUE   EA      ID        PART ID      USED ON
IN REPAIR    NA
                                        4789CW   9981NT          4801Z

STOCK CATEGORY S   ANNUAL USE        45

  CONTROL CODE C   UNIT COST        4.75

    REPAIRABLE N   LEAD TIME(DAYS) 30

LAST REVIEW 8404   REORDER POINT     8

                   ORDER QUALITY     7

         Enter PART ID, or "Q" to exit.
```

Figure 4-3   Simplified Part Record with Availability

Note that the field notations are those commonly used in computer systems analysis. "X" indicates alphanumeric contents and the number in parentheses is the number of digits. For example, an X (12) part number could be P0123456789A. A "9" field means that only numeric entries are acceptable, which allows additional computer editing and data positioning that are not available with X fields. The letter Z used in a numeric field starting with the leftmost digit indicates that if those actual digits are zeros, they should be eliminated and replaced with spaces. Thus a 9999 field would show 0082 but a ZZZ9 field would display 82 preceded by two spaces, making it easier for a person to read. Field sizes and the number of significant digits are determined by the size of numbers used by the user. The (S) means scroll fields, which can be rolled over to contain as many lines as necessary. Figure 4-2 shows the data displayed on a VDT screen. Figure 4-3 illustrates a simpler format that is suitable for small service organizations, and also includes the functions of location and quantity.

Keep only necessary data. These database fields should cover any event that may occur. If, however, the data are not needed, then eliminate that element from consideration and do not waste the time and effort required to gain and keep accurate information. It is even worse to have false information than it is to have none at all. Information can be expensive. The benefits of having information must be weighed against the cost.

5

Cataloging Practices

MEDIA

THE MEDIUM USED FOR cataloging parts probably will be paper for years to come, with a gradual transition to microfiche and computerized electronic data retrieval. Parts data have historically been written on cards or printed on paper, often in the format shown in Table 5-1.

Sequenced in part number order or alphabetically by noun and description, this type of catalog is useful for an industrial parts crib or field service stockroom. As the volume of paper

Table 5-1
Parts Data

Part Number	Description		Location	Unit of Issue	Dealer Price	Unit Price
12345	Fuse	,15 AMP,SLOW BLOW	A12B1	EA	1.04	1.25
12346	Relay	,20 AMP,DPDT,TI DELAY	E01A3	EA	6.25	7.80
12349	Switch	,20 AMP,SPST,FOR COLEMAN X21	B09E8	EA	2.96	3.55

increases, inventory systems specialists usually recommend the use of microimaging. Microfilm that contains a 35-mm photograph of each page in sequential order is handy for high-volume storage, but microfiche that can store 80 pages per 4×5 film is more efficiently distributed to many users, and the desired information is more easily retrieved from the flat fiche. Where many field representatives are involved, the cost of mailing alone will pay for the difference in cost between paper copies and the film.

Film image is excellent for typed single-line entries such as parts catalogs, but is not as good for schematics and detail that need to be closely traced.

Revising and updating parts information is a continual challenge. If change sheets and additions are distributed, the receiving people rarely note all the changes on their base documents. Since the objective is to provide accurate information to the end user with minimum effort, complete pages, or even whole documents, should be distributed whenever possible.

Another concern is having information readily available in case a computer-based system is interrupted. It is recommended that a complete catalog be printed about every three months, or whenever significant changes occur. Catalog listings should be ordered in the several sequences most useful to parts requesters: Part Number, Noun Description, and possibly Equipment Used On.

WAYS TO IDENTIFY PARTS

Useful ways in which a part may be identified in a complete system include:

1. Organizational part number
2. Vendor's part number
3. Manufacturers' part number
4. Word description by noun and modifiers
5. List of parts for a commodity code
6. List of parts used on a specific equipment or a product line
7. List of parts made by a specific manufacturer
8. List of parts purchased from a specilfic supplier

The key information in most systems is the internal part number, which, as previously described, should indicate a unique stockkeeping unit (SKU). All other ways of finding a part should reference the part number.

Persons trying to identify a specific part will usually ask for it in descriptive terms, "I need those little rubber washers and springs that fit inside my single-handle kitchen faucet," or "I need an oil filter for a 1984 Dodge pickup truck." Here the query will best be by noun nomenclature for a "WASHER, KIT W/SPR & TOOL, FOR DELTA SGL-HDL," or "FILTER, OIL, 1/4T TRK, CHRYSLER." The noun nomenclature is related to commodity code as described in Chapter 4. It is even possible to have a computer program that identifies the possible commodity codes for each part description and then allows the operator to decide which is best. Wire nuts, for example, that are used to make electrical connections would be "NUT, ELEC, 10 GA-2 WR." Since the search check for commodity code against the noun would be only against the word "nut," it would probably show the choice to be mechanical or electrical and the user would select the appropriate code for the group and class. Other ways to find the oil filter would be to do a computer search of all parts stocked for trucks, or all parts manufactured by Chrysler, or all parts purchased from NAPA. A typical catalog page is illustrated in Figure 5-1.

INFORMATION CONSIDERATIONS

There are ten considerations regarding information systems that should be addressed to assure success. They are shown in Table 5-2.

Table 5-2
Information Considerations

Reliability	Excess
Validity	Redundancy
Relevance	Flexibility
Quality	Timeliness
Insufficiency	Cost

```
P A R T S   C A T A L O G   B Y   D E S C R I P T I O N        PAGE:   1

STOCK NUMBER        NOUN        DESCRIPTION

       201          ABRASIVE    ,SILICONE,CARBIDE,120:15 DEG
       202                      ,SILICONE,CARBIDE,120:30 DEG
       208                      ,SILICONE,CARBIDE,120:45 DEG
       209                      ,SILICONE,CARBIDE,120:60 DEG
       218                      ,LEAVITT LAPPING MACHINE
        49          ABSORBER    ,SHOCK,MILLER,MANYARD,SHOCK
                                6FT SPECIAL:SECURITY/SAFETY/FIRE PROTECTION,
                                *SPECIAL PURCHASE-SEE I&E FOREMAN BEFORE
                                  REORDER
        50                      ,SHOCK,TEFLON
     29518                      ,BALSBOUGH LAB MONITORS,CELLS,BOXES
        81          ACCELEROME,ENDEVCO
        82                      SRVO POSITION MONITORING, WILCOX 3BC
      3002          BEARING     ,ROLLER,BALL,STNLS,3.6MM
     24355                      ,ROLLER,STNLS,W/SEAL,3"
     99652                      ,ROLLER,PHOSBRONZ,1/2"
    123456                      ,ROLLER,PHOSBRONZ,3/4"
```

Figure 5-1 Typical Parts Catalog by Description

Reliability means that the same measurement will be obtained every time, given the same circumstances. Validity says that the numbers mean what we think they mean and that they apply to our specific situation. It is related to relevance, which means that there is high correlation with the real world. Quality refers to the accuracy of the data and how pure the data are. If errors can creep in undetected, or if personnel forget to enter issues and receipts without being caught, then quality will be low. Insufficiency means there are not enough data to answer the question accurately, which is often the case with low-use parts. On the other hand, there can be too much data and we can be overwhelmed with excess. Information systems are often data-rich and information-poor. Redundancy occurs when the same data are gathered from several different sources, or the same objectives can be met in several different ways. The saying, "A person with two watches is only confused," has relevance to getting different parts data from different sources. It is better to rely on one source and concentrate efforts toward making the data the best possible. Flexibility does help in case one scheme fails and alternative approaches are necessary.

Timeliness refers to the fact that the value of information decreases with age. A computer printout, for example, was only valid the instant the data it refers to were produced. If there is relatively little change, that may be no problem. However, if there is frequent change, then interactive real-time data are much more valuable. The cost of gathering all this information naturally must be considered. Knowledge is a valuable commodity, especially when dealing with parts, and the benefits should outweigh the costs.

MANUAL VERSUS COMPUTER

Several references have already been made to the use of computer systems. While the computer, which will be covered in more detail in Chapter 16, provides fast, efficient, accurate processing of large amounts of data, this does not mean that manual records are defective. In fact, even with a computer sys-

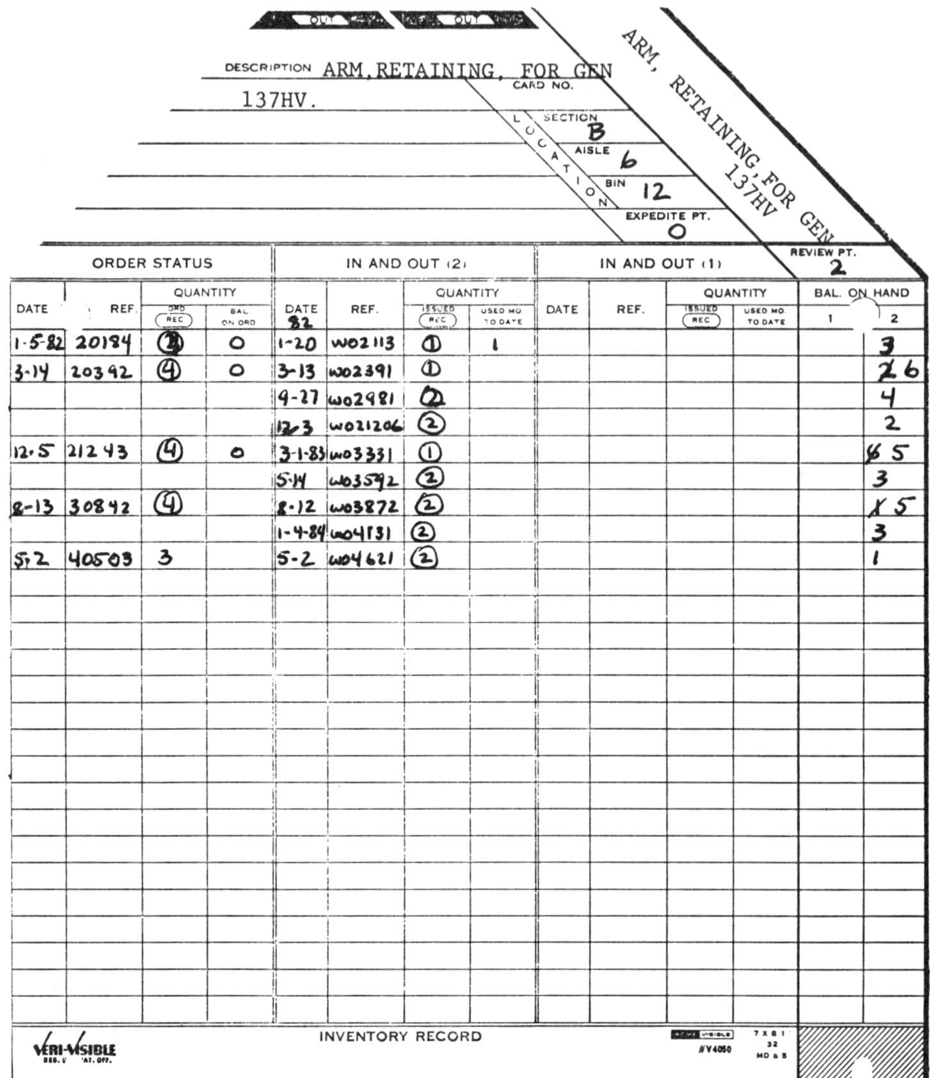

ORDER STATUS				IN AND OUT (2)				IN AND OUT (1)				REVIEW PT. 2
		QUANTITY				QUANTITY				QUANTITY		BAL. ON HAND
DATE	REF.	ORD (REC)	BAL. ON ORD	DATE 82	REF.	ISSUED (REC)	USED MO. TO DATE	DATE	REF.	ISSUED (REC)	USED MO. TO DATE	1	2
1-5-82	20184	(2)	O	1-20	W02113	(1)	1						3
3-14	20392	(4)	O	3-13	W02391	(1)							26
				9-27	W02981	(2)							4
				12-3	W021206	(2)							2
12-5	21243	(4)	O	3-1-83	W03331	(1)							65
				5-14	W03592	(2)							3
8-13	30842	(4)		8-12	W03872	(2)							5
				1-4-84	W04131	(2)							3
5-2	40503	3		5-2	W04621	(2)							1

- Inventory Records
- Obsolescence Control
- Automatic Review Point

Figure 5-2A   Acme Veri-Visible© System—Card 1

DESCRIPTION

ARM, RETAINING, FOR GEN 137HV

REVIEW PT.	2
PROCUREMENT TIME	20 da
MINIMUM ECON. BUY QUANTITY	3
UNIT OF PURCHASE	Ea

USAGE

YEAR	JAN.	FEB.	MAR.	APR.	MAY	JUNE	JULY	AUG.	SEPT.	OCT.	NOV.	DEC.	TOTAL
1982	1		1						2			2	6
1983			1		2			2					5
1984	2				2								
19													
19													
19													

VENDORS

1	GENEROUS ELECTRIC, 500 W. HOT SHOT, LIGHTNING, KN	(413)354-2566
2	RELIABLE POWER, 12 BRIGHT LIGHT PL, EDISON, NJ	(288)412-5344
3	POWER PRODUCTS, POWER LANE, UPPER VOLTA, NY	(716)344-2000
4		
5		

ON HAND / ON ORDER	DATE / REQ. BY	QUAN. WANTED	DELIVERY SCHED. QUAN.	DELIVERY SCHED. DATE	APPROVALS	LINE #	DATE / P.O. NO.	VEN	ROUTE / F.O.B.	BUYER / ACCT. NO.	PRICE / UNIT	DATE PROM. / DATE COMP.
0	1-5-82	4	4	1-27	NCF	1	1-5-82	2	Adkin	BND	48.50	1-30
0	1-30-82						20184		Trk		Ea	1-27
2	3-14	4	4	4-1	NCF	2	3-14-82	2	Trk	BND	48.50	4-1
0	4-1						20392				Ea	4-13
2	1-5-83	4	4	1-5	NCF	3	12-15-82	2	Trk	BND	48.50	1-5-83
0							21243				Ea	1-4
1	9-2	4	4	9-2	NCF	4	8-13-8	2	JRK	KPC	53.00	9-2
0							30842				Ea	9-3
1	5-20			5-20	NCF	5	5-2-84	2	TRK	BCP	58.00	5-20
3							40503				EA	5-20

PURCHASE HISTORY - TRAVELING REQUISITION

VERI-VISIBLE REG. U.S. PP. ACME VISIBLE · V4051

- Supervisor Review
- Complete Purchase History
- Traveling Requisition

Figure 5-2B   Acme Veri-Visible© System—Card 2

tem, the more difficult challenges in service parts management are the physical and human ones. Several proved manual systems exist. It makes more sense to use one of them instead of trying to develop your own. The Acme Veri-Visible® system is illustrated in Figure 5-2.

The Veri-Visible System consists of two cards. One is the "Part Record and Quantity" card, which always stays in alphabetic order in the card bin. The second is the "Traveling Requisition." When a stock part is needed, the traveling requisition card is pulled from the bin and used to provide the purchasing information. When the order is received, the traveling requisition card is returned to its position. Since the part number is the key information on a manual parts system, finding the part depends on knowing that number. A good cross-reference is the strip-chart catalog illustrated in Figure 5-3.

The strips are 8½ inches long and about one-half inch high on card stock. A typist enters the part number and noun description. It may include price, where used, and the supplier. If two strips are typed for each part, then one can be arranged in the frame in part number order and the other can be in a separate frame in alphabetic part noun description order. If worth the effort, additional strips could be arranged by "where used" and/or by manufacturer. Copies of the catalog can be made simply by placing the removable "pages" on a copy machine. As new parts are added, their strips may be inserted in the proper location on the frames.

Accuracy of the records and the maintaining of perpetual paper or computer information consistent with physical realities are vital. If the data-collection methods are awkward and time consuming, the data will not be collected. An uncontrolled stockroom, meaning that anybody can take the parts, rarely will have accurate information on parts status. Someone must be in charge of assuring that the right parts are on hand, and a good record system provides the necessary tool. If it is not possible to have a video display tube (VDT) at the counter and enter data immediately for every part, then a list in which one writes down every part number with quantity and work order reference can be used. Pick tags can be attached to individual parts, and then pulled off and put in a basket for separate logging. A second box containing the safety stock may be in the bin with a reorder card attached so that when the main supply is exhausted, the safety stock box can be opened and the reorder card simply dropped into the purchasing agent's basket. Parts that are hung on a pegboard hook can utilize the same principle, with a reorder tag that is suspended on the rod at the point where the reorder quantity of items will remain after the reorder card is moved.

It is important that a complete audit trail be maintained on every important part, from the time someone needs it until that part is used. Checks and balances should be provided to assure that all ordered parts are, in fact, received; that they are paid for on time, and only once; and that they are effectively used for the purpose for which they were intended. A parts record system should include audit controls and safeguards that preclude dishonest actions from even being attempted. For example, one person may request and order parts, a separate person should confirm their receipt, and a third person should authorize the payment. Cycle counts of physical versus perpetual inventory should be conducted with emphasis on essential high-cost parts that have uses outside the business and therefore might be subject to disappearance. To provide a factual basis for service parts management, the information must meet the Table 5-2 criteria whether in paper or in computer form.

```
- - - - - - - - - - - - - - - - - - - - - - - - - - - - - - - - - - - - - - - -
   12345 - POWERSUPPL, 500 W, 115 V, HEAVY DUTY, FOR CONTROLLER HVX333
- - - - - - - - - - - - - - - - - - - - - - - - - - - - - - - - - - - - - - - -
   POWERSUPPL, 500 W, 115 V, HEAVY DUTY, FOR CONTROLLER HVX333 - 12345
- - - - - - - - - - - - - - - - - - - - - - - - - - - - - - - - - - - - - - - -
```

Figure 5-3   Strip Chart Examples

6

Provisioning

SERVICE PARTS SELECTION

Now THAT WE HAVE determined what records need to be kept on parts, we need to decide which parts we need to keep those records on. Determination of these parts is the subject of this chapter. Deciding how many is covered in Chapter 8 on forecasting.

The need for provisioning of any parts is driven by requirements for parts replacement due to breakage, consumption, or wearout. The first question to ask is, "What parts are likely to fail?" That query is best directed to engineering and field personnel. Data can come from comparable other products, laboratory tests, competitive products, quality-assurance trials, vendors' recommendations, and military handbooks. MIL-HDBK-217, *Reliability Predictions for Electronic Devices*, provides generic part failure rates that can be used to predict failures. Experienced field technical personnel know that devices such as relays, switches, lamps, motors, and power supplies are prone to failure. Any devices that require preventive maintenance change, such as automobile oil, filters, antifreeze, and wiper blades, should be stocked.

The objective of provisioning is to preplan as much as possible what parts should be carried on the shelf so that they will be there when needed. Certainly other parts, often called "nonstock" or "direct to order (DTO)," may be obtained, but not as rapidly or as efficiently.

The histories of similar equipment are the most effective guide as to which parts to supply. That is so because the environment and personnel operating and maintaining the equipment have such a great influence. Often in doing mathematical predictions and laboratory tests to predict field performance, the total failures will be correct, but for different reasons than those predicted. Some parts fail that were not expected to fail and other failures that were forecast do not happen. Under any conditions there is probably a translation multiplier, often called a K factor, of two or three times more failures in the field than there were in the laboratory. Many of these will be caused by newly trained, inept personnel who are learning how far equipment can be stressed. It is very difficult to accurately simulate the dirt buildup, electrical variations, and other diverse challenges encountered in the field, so some new failures may occur. Don't forget that some parts are used to support laboratory tests and then training, so those requirements must be considered. Note that provisioning gets the process started, but stocking lists must be dynamic and adjusted as experience is gained.

ESSENTIALITY

The concept of essentiality greatly aids determination of what parts should be stocked. A part that has safety implications on major facilities or equipment should be rated essentiality class 1. Parts that are critical to the operation of major facilities and equipment should be class 2. Without them, that equipment would be down and revenue loss would be great. Class 3 parts contribute to degraded operation, as in cams or snubbers that wear and begin to cause jams and lower production rates. Class 4 parts are cosmetic, such as knobs and labels. To relate the parts need to the maintenance priority, a class 1 or 2 part need will generally be a special "drop everything and run" emergency service call. Class 3 parts would be used on work scheduled in the normal work scheme. Class 4 parts would be installed the next time a person is at the equipment and can do the job conveniently. These four classes are summarized in Table 6-1.

Table 6-1
Essentiality

Essentiality	Description
1	Safety, or causes secondary failures
2	Equipment down and losing revenues
3	Degraded operation, limping along in reduced fashion
4	Cosmetic, when convenient—minor parts, tools, test equipment, and other nonequipment parts

The military has rules for provisioning that say, "If you use one or more of a part every three months, it should be stocked at your location. If you use one every six months, it should be stocked at the next higher location." These rules, however, disregard essentiality, lead time, and the cost of not having the part, which are important considerations. A better rule for determining whether a part should be stocked is: If the expected value of having that part exceeds the carrying cost for that part, then it should be stocked. The carrying cost can be calculated as discussed later. The expected value requires more management judgment.

Expected value comes in two ways:

1. Profit on parts available to be sold.

2. Cost avoidance for parts that are necessary to do contract work and would require expediting and additional costs if they were not available. Every for-profit business should expect to earn at least as much profit on parts as on labor. If the desired profit margin on operations is 30 percent, for example, then that factor should be included in the calculation, as shown in the following equation.

Part Cost of Supply versus Expected Value of Demand

(Item cost × Holding % × Return on assets × Quantity) ≤
  (Probability of demand × Value gain of loss Prevented)
For example:

($10 × 0.30 × 1.3 × 1) ≤ (0.2 × $27)

or

$3.90 ≤ $5.40

Since the cost of carrying the part is less than the cost of not having it, then stock it.

The probability of need will be covered further in Chapter 7 on statistics. It is, however, largely dependent on educated judgment. When we need a part, our concern is usually whether we have one or zero parts in the bin. An excess beyond requirements doesn't make any difference to a requestor. The challenge is how to assure that we do have the needed part.

STANDARDIZATION

Provisioning personnel must start in the early phases of product planning to stimulate use of common parts throughout as many products as possible. Design engineers generally have little appreciation or training or direction in the value of standardization. There will often be an organization called Components Engineering that is charged with finding, testing, and evaluating the various alternatives for components to achieve a specific function. It may also provide vendor qualification once the choice is narrowed. Even in a factory or government facility, where most equipment is being purchased rather than internally designed, commonality is important. Does the manufacturer provide a complete list of spare parts? Are similar parts already available in inventory? What will be the expense of stocking those additional spares? Can existing spares be utilized? Are any of those spares high-failure, unique parts that will be very expensive over the life cycle? Can our personnel fix any repairable parts? If not, where must they be sent and how much will it cost? What is the system for providing replacement parts? Remember that over a product's life cycle, replacement parts may add up to a large percentage of the total cost. The time to pay attention is during the early phases of design or before the product is purchased.

KITS

Consolidating all the parts needed for standard specific types of work makes good sense. These prepackaged materials should be prepared for preventive maintenance, modifications, and standard repairs. Everything necessary to do the job should be in the kit, including a list of kit contents, installation instructions, clean-up materials, any reporting forms, and even special tools. One of the major objectives of kits is to help parts personnel and users make effective use of their time. Chasing parts can be a major time waster. A PM kit for any dirty task should include, for example, a disposable drop cloth that can be spread on the floor surrounding the work area, and a plastic bag in which to dispose of dirty materials. Any repairable parts should be packaged with a repair tag and return packaging, right down to the sealing tape and mailing label. Depending on the value, any excess parts that were in the kit, but left over from the job, normally should be returned to the stockroom and restocked under their individual part numbers. If only a few washers are left and their value is less than the effort necessary to return them, then they should be discarded. If the value exceeds the labor cost to restock them, or if they are in short supply, then they should be returned to stock.

PACKAGING

The objective of packaging for service parts is to provide protection, ease of handling, and identification. When deciding on packaging, put yourself in the position of the stock clerk who picks or shelves the items, the packer who prepares shipments, and the mechanics and technicians who use them. Wherever possible, individual line items should be packaged for the end user. If the end user is a factory mechanic, plumber, or heavy equipment dealer, the parts may need no packaging and may be stored in an open bin. Parts for field personnel should be packed in plastic envelopes, cardboard tubes, or foam-filled boxes, with particular attention paid to avoiding transportation damage. Shrink wrapping is good for small parts. Small fasteners and hardware, for example, can be placed on a piece of cardboard that contains their description and even installation instructions, with clear plastic shrunk over them. In that way they can be easily viewed to confirm that they are the right items and will be easily identified. The many prepackaged items for sale in hardware stores today illustrate this type of packaging. In most cases the technician will use one piece to replace a single piece in the field. This will occur whether the replacement is done for PM or for emergency reasons. It makes sense then to package items one piece to a package or one set to a package. This policy requires close coordination with manufacturers and suppliers, and may even require a separate packaging function within the service parts organization.

For electronic parts that may be damaged by static electricity, antistatic bags can be sealed with a tamper-evident label. When the label is removed, the word "Opened" appears on the label and on the bag's surface. Another handy method of packaging, especially where a central stockroom prepares parts orders for many jobs, is by using small plastic bags. These bags may be of the ziplock type or ones that can easily be stapled or taped closed. A label with the part number and description and quantity on it should be fastened to the outside. Some operations print gum labels on a computer printer at the same time the parts pick list is printed. The common mistake to be avoided is putting many different parts in the same bag. The author once saw a large bag covered with gummed identifier labels and containing over 200 items of small hardware. This was sent to a field tech rep who would have to place each of those parts carefully into the bins of a carrying box. It is not surprising that the rep sent the package up through management channels with proper ceremony commemorating the parts packer who perpetrated the travesty. While it would have been a little more work for the packer to place each of these parts in separate bags by line-item part number, it would have paid for itself many times over in efficiency at the receiving end where the tech rep would know what parts had been received and could put them directly into the individual storage boxes.

SUPPORT PHASES

The major phases of product-part support are as follows:

1. Preproduction for tests, training, and field trials
2. Product introduction
3. Normal life
4. End of production
5. End of life

Figure 6-1 illustrates these phases. Each of these phases has unique characteristics in the requirements for parts. The question of what parts should be stocked is much more difficult to answer in the preproduction and product introduction phases and then becomes generally routine during normal production and use. It will again become a concern during end of life because wearout may begin to increase failures; however, experience with this effect should have been gained by the time the information is required.

Pre-Production	Intro-duction	Normal Life	End Production	End Life

Figure 6-1   Product Support Phases

Preproduction

Service support for new products of a special or unique nature is not guided or controlled by any logistics system presently available. Most high-technology organizations develop new products in several sequential steps—in effect giving birth to a concept that must be nurtured through stages of sitting up, crawling, walking, and finally running. These stages are generally necessary to develop, analyze, and improve a proposed product to meet both technical and marketing goals.

Market research surveys, panels, and simulations are valuable tools for proving acceptability of products, but the major gain in information is possible only by putting actual equipment in field locations for customers to use under observation. On a typical new project, these field tests may begin with breadboard machines and include engineering models, prototypes and preproduction hardware. Preproduction parts are required for support of those machines that are few in number, relatively unproven, and functional for a predictable time period, after which they are scrapped.

Equipment. "Typical" machines will be complex, expensive electromechanical devices. Breadboard operating models are very expensive and are designed to have flexible components that can be redesigned and altered frequently to discover what parameters must be controlled. Nearly all the equipment's components are handmade. They are strictly laboratory devices to be operated by technicians.

Engineering models are primarily intended as the first equipment in a form similar to the expected final product. They are used to refine the controlling parameters to manufacturable limits. These machines will have panels and design compatible with the final intended environment, however, they probably will not meet all the performance goals set for the production product. They are individually handmade and must still be operated by technicians.

Only a few prototypes will be made by the engineering department to confirm design, so the preproduction machines that follow closely are usually the next available for test. While these machines probably will require further changes before production, they should be built by manufacturing engineers using the same tools, gages, and techniques that, it is hoped, can be used by manufacturing personnel in full production.

The majority of the parts are frames, wiring, and common securing hardware that should never break or wear out. Some parts, of course, have a probability greater than zero of causing system degradation at some time during the machine's life. There will be very little objective data on component usage at the time the field test starts.

Materials that are consumed in the process are designated "consumables." The parts that may fail and must be repaired or replaced are designated field replaceable units (FRUs). Analysis of machine characteristics reveals the following conditions that require a unique logistics system.

1. Most parts are individually made by model makers and are comparatively expensive.

2. Most parts in a subsystem are issued as a package.

3. All spares anticipated to be needed should be ordered at the same time as initial build parts. This helps to ensure commonality of parts. It also realizes the lowest possible production cost since the model makers will do several parts at the same time faster and more efficiently than they can do the same number at different times. However, there is no best economic re-order quantity.

4. The length of time a prototype machine will be in use can be determined at the time it is built.

5. Test machines are written off as expense. When tests are finished, the machine will probably be scrapped and unused spares will have little or no value.

6. Spares support for test units installed at the customer's site can be provided via an "initial spares kit." This kit should be shipped with the equipment and provide the necessary on-site initial support. Cost of the initial spares kit should be absorbed as part of the test unit support cost.

Personnel. Let us assume for illustration that production machines eventually will be run and routinely maintained by operators. Technical representatives will provide repair and maintenance beyond the operators' capabilities. They will perform in a territory controlled by branches that can furnish management and logistical support.

Breadboard and engineering models, however, must be operated by technical service aides because of the machine's complexity, variability, need for frequent repair, and requirements for sophisticated data collection.

Problems. There are three basic problems to be solved.

1. Location of spare parts
2. Quantity of spare parts
3. Cost of spare parts

Model. These three problems are, of course, interrelated. The approach is to solve the location and quantity simultaneously and then determine total costs. A computer is available for operating the working model. The initial machines will have tech service aides operating them so the person on the scene is qualified to repair any part. The next level of support is the factory; see Figure 6-2.

We have M machines and are examining whether the spares should be produced on an "as needed" basis at the factory or inventoried with machines. If we use the factory as the source, the expected cost for supplying the part is $C_{ri} = C_{pi} + C_{mi}$, where C_{pi} is the purchase or manufacturing cost and C_{mi} is the remote

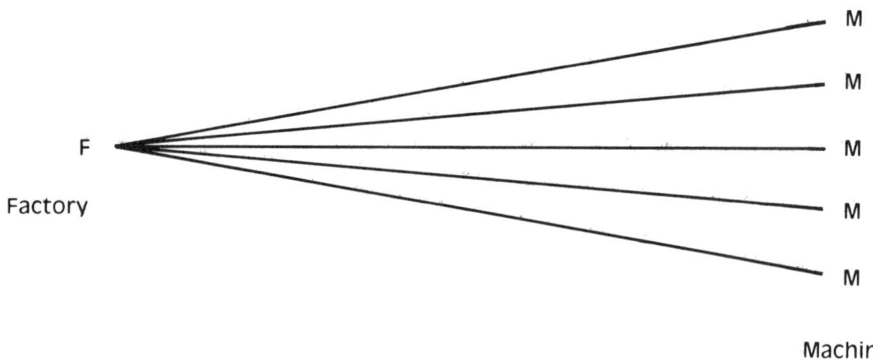

Figure 6-2   Direct Factory Support

supply cost, including lost revenue, customer goodwill value lost, and expediting costs.

If we choose the option of inventorying parts with the machine, the expected cost per repaired failure is

$$C'_{ri} = C_{pi} + P_{si} + C_{mi} + MR_{ni}C_{pi}$$

where C_{pi} is the same purchase or manufacturing cost; P_{si} is the probability of stocking out of the part; C_{mi} is again the remote supply cost, including revenue and goodwill loss and expediting; M is the number of machines; R is the cost of carrying the inventory in terms of money rates, taxes, insurance, and handling and storage costs; and n is the quantity of the part kept in each inventory stock.

Labor for repair is treated as a sunk cost and disregarded. There is no term for scrap or part wastage because we expect to use up the stock during the products' life. The choice of part location will be made on the basis of whether $C_{ri} > C'_{ri}$. The location with the lowest expected replacement cost will be used.

The function $C'_{ri} = C_{pi} + P_{si}C_{mi} + MR_{ni}C_{pi}$ will have a minimum cost point on ni, which gives the value to be compared with C_{ri}. The computation of C'_{ri} will be done in stepwise fashion assuming $ni = 0, 1, 2, \ldots$ until a minimum point is reached. $ni = 0$ means supply only from the higher factory level.

The tricky part of the computation is estimating P_{si} as a function of ni. The normal supply system can replace parts used in seven days at no additional cost. We express this period of resupply as T years and assume the arrival of failures as Poisson:

$$P_{si} = \sum_{K > n_i}^{\infty} \frac{(rt)^K e^{-rt}}{K!}$$

where

$r = ri\ mi$
$k =$ total failures

Operating Values. The determination of realistic values is one of the most difficult areas. Through work with product design and development engineers, they agree to provide the following information on each numbered part i:

> PNEN—Part number English name
> PCST—Part cost
> PEUY—Part expected usage per machine year
> PEFF—Part essentiality
> OXPS—Other part substitutable
> RLVL—Replacement level (operator or tech rep)
> RORS—Repair or scrap (TR, factory, or scrap)

Other data elements necessary are

> TMRS—Total machines requiring service
> YMIU—Years machines in use
> CCRP—Carrying cost R percent
> RSCC—Remote supply cost C_{pi}

The opportunity loss is dependent upon essentiality. Marketing experience and judgment might determine cost relationships as shown in Table 6-2 where the equipment is leased.

Table 6-2
Opportunity Loss Based on Essentiality

Essentiality Level	Opportunity Loss in Dollars per Hour t
1	$25\ t + 35\ [0.25\ t + 0.05\ (t-1)]$
2	$\dfrac{25\ t + 35\ [0.25\ t + 0.05\ (t-1)]}{2}$
3 or 4	Negligible up to seven days, during which time the part should be easily supplied.

This assumes loss of potential \$35 contribution income for essentiality 1 of 25 percent the first hour, increasing 5 percent each additional hour t in the form:

$$\text{Loss} = \text{Goodwill }\$25\ t + \text{Foregone income }\$35\ \frac{[25\ t + 5(t-1)]}{100}$$

Very high machine base rent is the reason for the high \$25 goodwill estimate. The revenue loss is based on expected average \$35 per hour. If machine is down for one hour, about 75 percent of the work will be saved and run when the machine is repaired and only 25 percent of the contribution margin actually lost. As the machine is down for longer time periods, proportionately more potential income will be lost. For a level 2 part, the effect is assumed to be only half that of a level 1.

For the most frequent cumulative times, the losses would be as shown in Table 6-3.

Table 6-3
Revenue Losses

	Essentiality Level		
Time—Work Hours	1	2	3 or 4
1	$ 33.75	$ 16.88	0
2	69.25	34.63	
3	104.75	52.38	
4	140.25	70.13	
8	283.25	141.63	
24	850.00	425.00	
40	$1472.35	$736.13	0

The time to resupply is a function of the source location, with expected times based on actual field data for existing products. See Table 6-4.

Table 6-4
Time to Resupply

Stock Location	Time to Supply to Machine, Work Hours
Machine	0
Tech rep	3
Team	4
Branch	8
Factory	24

Only the machine and factory locations are feasible with a small number of test machines. It can be easily seen from the matrices that loss due to a level 1 part that must be supplied from the factory will be $850, and to level 2, $425.

The total number of machines requiring service and the number of years these machines will require service are determined from the product marketing plan. In this case carrying costs are estimated by financial personnel to be about 25 percent of the part's value per year. Expediting costs are not separated because the time loss avoided by expediting will compensate at least for overtime worked, special delivery to the airport, and so on. The time loss values can be used to determine the amount of special expediting that should be given each needed part.

Sample Computations.   Combining the operating values and the model, we pick a typical part. This programmer costs $100, is essentiality level 1, and is expected to need replacement three times each machine-year. It is repairable by the factory. Replacement of used parts is every seven days, or 0.019 year. There are five machines to be serviced for one year. Thus, we have the values:

$$C_{pi} = 100 \qquad R = 0.25$$
$$C_{mi} = 850 \qquad T = 0.019$$
$$M = 5 \qquad r = 3.0$$

It is now possible to solve

$$C'_{ri} = C_{pi} + P_{si}C_{mi} + MR_{ni}C_{pi}$$

doing first

$$P_{si} = \sum_{K > n_i}^{\infty} \frac{(rt)^K e^{-rt}}{K!}$$

Hand solution is easiest if each factor is set down as it is determined. Since there is one part in the machine to fail at rate r, the computations of total failure k begins with 2. See Table 6-5.

Table 6-5
Computations of Total Failure

K	$(rt)^K$	e^{-rt}	$\dfrac{(rt)^K e^{-rt}}{K!}$	Σ
2	0.003250	0.944	0.003060	0.003060
3	0.000185	0.944	0.000175	0.003235
4	0.000014	0.944	0.000005	0.003240

Obviously the exponential effect of k causes the expected P to stabilize for practical purposes very quickly.

Table 6-6 shows the C'_{ri} computations.

Table 6-6
Total Cost Calculations

n	C_{pi}	+	$P_{si}C_{mi}$	+	$MR_{ni}C_{pi}$	=	C'_{ri}
0	100		850		0		950
1	100		3		125		228
2	100		0.25		250		350.25
3	100		0.02		375		475.02

The minimum expected cost of supplying a needed part is $228, which occurs when five spare programmers are made and stocked one at each machine. Thus this is the answer to Problems 1 and 2.

For the programmer we should make the initial five spares at the same time the machine build parts are made. This will require an initial outlay of $500. When a programmer fails and is replaced with the spare, the failed programmer will be immediately returned to headquarters along with the order for a new one. It would be repaired and returned within seven days.

The time required to do these calculations by hand and the logical method of solution for each of about 1000 parts dictate use of a computer for working solutions.

Total Parts Costs. The total amount of money required to suport spare parts for a year will be

$$C_t = M \sum_i^\infty riC_{pi}$$

for our sample programmer, assuming C_{pi} equals repair or replace cost.

$$C = 5 \times 3 \times 100 = \$1500 \text{ per year}$$

The principal source of error is the estimate of usage rate r. Since there can be as many as 1000 different parts replaceable, there should be a great deal of error smoothing. If we have 1000 parts and assume that all cost the same and have a random relative standard error of 100 percent and no other source of error,

$$C_t \text{ relative error} = \frac{100}{\sqrt{1000}} \approx 3\%$$

Even if the failure rate data were perfect, the number of failures actually occurring in a fixed period, such as one year, is still subject to chance fluctuations. To eliminate excess parts production that would have little or no useful value, initial part build should be the smallest number of parts consistent with the service objectives discussed under Problems 1 and 2. We have assumed no penalty for making later production runs.

Accurate records of parts usage and failure rate must be maintained so the system can be analyzed and updated periodically. Every time a demand is placed for a part, the data, it is hoped, can be reviewed to determine whether the demand is normal and what action, if any, should be taken to improve logistic support.

Toward the end of the operating period, it will be desirable to reduce the inventory of parts and let the system wind down. For a few very cheap, high- (low-number) essentiality parts, it may be desirable to keep stocks through the end, since it may be cheaper to throw away the leftover part than to have to expedite one from the factory. This decision should be made for every part stocked by substituting $r = 1.00$ in the C'_{ri} calculations.

Money is always a constraint. The possibility always exists that not enough money will be available to have everything we want. Cash flow, too, occasionally may dictate that fewer dollars be spent early even though it is expected to cost more later. To be prepared for operating budget analysis, the printout of parts can be arranged for ease of evaluation. After computations are performed for location and quantity of parts, the total cost of those initial parts should be computed. Then the total expected usage and cost for the service period should be set down. A printout or VDT display in part number sequence is necessary for reference. It should follow the form of Table 6-7. Grand totals should be added for initial spares cost and year's spares costs.

Table 6-7
Format for Initial Parts Data

```
999999 XXXXXXXXXXXXXXXXXXXX 99999 9 999999 X X XXXX 99 99999 999 9999999
  (1)          (2)           (3) (4) (5)  (6,7) (8) (9)(10) (11)   (12)

 (1)      Part number
 (2)      Part english name
 (3)      Part cost
 (4)      Essentiality
 (5)      Other substitutable part
 (6)      Replacement level
 (7)      Repair or scrap
 (8)      Spares location
 (9)      Spares initial quantity total
 (10)     Cost of initial spares in dollars
 (11)     Spares year quantity total
 (12)     Cost of years spares in dollars
```

The parts then should be grouped by essentiality levels and arranged by increasing unit part cost within each level. This gives us a display of parts in order of descending relative value. Since our earlier computations showed that essentiality class 1 parts have about twice the contribution of class 2 parts, a display in the class 2 array of twice the actual cost would make selection easy. If dollars were limited, the maximum value would be obtained by selecting initial parts in ascending order from the class 1 and class 2 lists up to the total money available.

As products reach higher volume and wider distribution, a more complex system should be brought into use. The next step should be as shown in Figure 6-3.

The model for this situation is basically the same as the two-location one, with the least-cost location selected. The probability of stock-out P becomes much more complex and must utilize the form:

$$P_{si} = \frac{n^s(i^n - s)e^{-n\lambda t}(i - s)!\sum\limits_{K=1}^{K=\infty} s(K,i-s)(\lambda t)^{s+k}}{(s+K)!}$$

Fortunately, the computer can provide rapid calculation of P_{si}.

As more information becomes available, the possibility of stocking parts at more than one location should be considered. If production time is very great on some parts, it might be best to produce a safety stock and have it available at a central location.

Repair should become possible as parts become more common. Models exist for this calculation and should be incorporated when data are available. Salvage value, too, is presently excluded, due primarily to lack of valid information and expected little effect. Calculations for determining service level were reviewed but are of little value if all products must be supported 100 percent. The ability to say, "We can supply X percent of our parts needs within Y hours at a cost of Z dollars," is not meaningful at this time.

Models have been developed on a firm practical foundation to permit objective determination of spare part location, quantity, and cost. It is expected that we will learn from the models

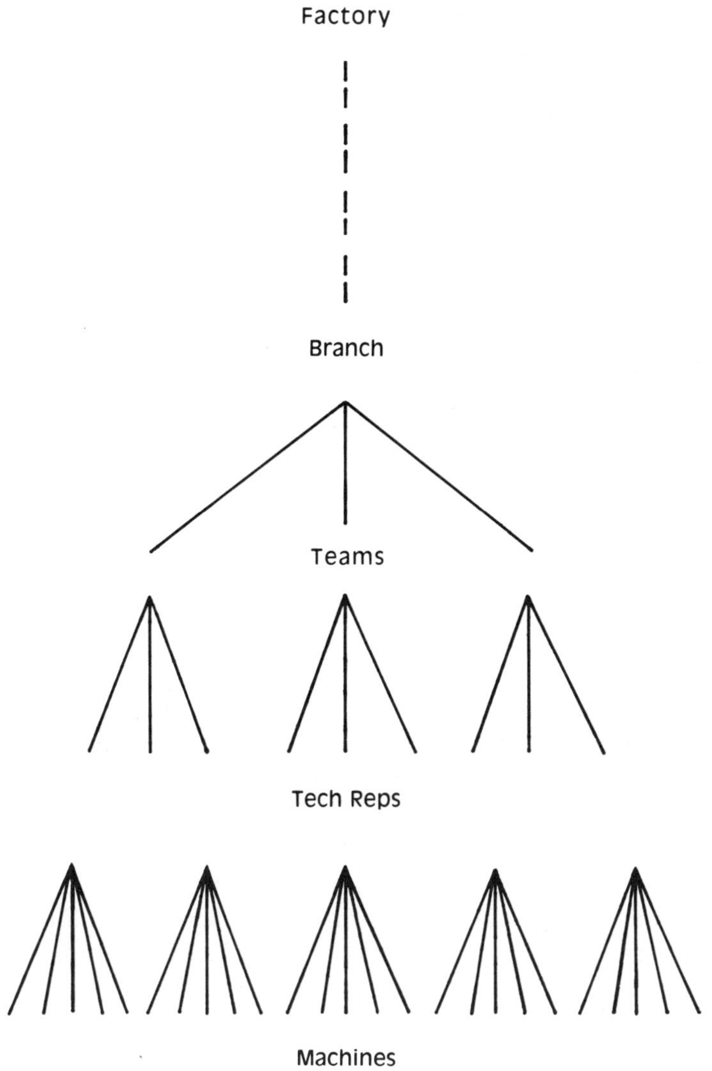

Figure 6-3   Repair or Resupply Hierarchy

so that rapid generalizations can be made in case of retrofits, part alterations, and other changes. The models should aid in establishing decision rules so that, for example, essentiality class 1 parts with expected usage of over ten per year and a unit cost of under $100 would be assigned to stock at the machine. In that way most spares decisions could be made individually at the time of new part issue and the complete program would need to be run only when major changes occur. The service level of local spares support should be examined closely during test unit operation. A complete log of all demands on the supply system should be kept. Demands filled from local stock should be logged also. The goal level for test units realistically should be 80 percent, with backup providing the remainder. It is rare that test units perform as predicted and a high engineering change rate complicates the issue.

Through this effort it is certainly possible to have improved spare parts logistics support for preproduction equipment.

Product Introduction

When a new product is introduced, either to field customers or within a factory, the concern is for a basic supply of replacement parts. Looking at it first from a factory perspective, reliance must be placed on the manufacturer of the equipment to provide a recommended spares list. Be cautious about simply accepting the list as published in the back of a catalog. Often such lists are prepared by a novice engineer during the rush of getting a manual ready for the marketplace, and have not had the benefit of experience. The best information will come from the field service personnel who maintain the equipment and from other users who have gained experience. Only the most essential parts that require a long lead time should be stocked.

Get a signed statement that the supplier will take back any parts that are in good condition if they have not been used after one year. In that way you will not be burdened with useless inventory. Figure 6-4 shows a typical document for parts returns.

From the supplier's side, particularly if large numbers of distributors or field locations must be stocked, filling that initial "pipeline" will use many parts. There should be a recom-

STEIGER TRACTOR, INC.

1101 1ST AVE. NORTH   P.O. BOX 6006   FARGO, ND   58108

DEALER RETURN GOODS AUTHORIZATION

Date ___________

Dealer Code ___________

Dealer Name ___________

Dealer Address ___________

Submitted by ___________

For Factory Use

Date Approved ___________

DRGA Number ___________

MATERIAL TO BE RETURNED — 'THIS REPORT IS TO BE USED FOR NEW MATERIAL ONLY SHIPPED OR INVOICED FROM STEIGER TRACTOR, INC. (MATERIAL FOR WARRANTY CONSIDERATION MUST BE PROCESSED ON WARRANTY FORMS.)

Receiving Report

Line No.	Quantity	Part Number	Description	Date of Shipment	Invoice Number	Net Each	Total	Qty. Appr'd.	Qty. Rejected	Comments
1										
2										
3										
4										
5										
6										
7										
8										
9										
10										
11										
12										

Why Material Returned ___________

Total ___________

BELOW FOR FACTORY USE ONLY

Date Report Received ___________

Comments ___________

Return authorized by ___________

Request to Return

☐ OK   ☐ Denied

Returned material checked by ___________

Credit Memo No. ___________

Date Completed ___________

White - Factory copy   Canary and Pink - Authorization (Must be included with returned goods)   Gold - Dealer's copy

Figure 6-4   Dealer Return Goods Authorization

mended stock list provided to all field locations and individuals. That list should be used as a guide, but also should be tempered by what the users perceive as necessary. In other words, the basic stock should be negotiated rather than dictated. But once it has been established, it should be controlled. Stockrooms and warehouses should be audited by management against the list to assure compliance. Any deviations should be corrected by either changing the list or modifying the parts and quantities on hand. Establish provisioning as a legitimate, straightforward process that is planned and controlled. Do not let it just happen!

Normal Life Support

At about six months after a new product is installed, early failures due to infant mortality should be eliminated and the equipment settled in for a long productive life. Preventive maintenance replacement parts can now be accurately planned. Experience will show whether the provisioning stock list was accurate, and will guide any changes. As will be covered later in this book, parts that are used in large quantities are relatively easy to forecast. The major challenge in service parts management is the parts with relatively low need, but the potential for causing big problems if they are not available. In reality this midlife support phase should involve many more product-years than any of the other phases.

End of Production

After a manufacturer has produced as many units of equipment as the market wants, the decision is usually made to stop production, and probably phase into another model. This situation will be looked at from two perspectives: (1) manufacturer and (2) user.

The manufacturer usually commits to support a product for a number of years after the last item is manufactured. This is typically seven years, although a few state ten years, and several say it is indefinite. How long that time period should be depends on the durability of the equipment and how long the major components should be expected to last, the tax implications, and the company's desired image. The seven-year period is a balance of depreciation time to give maximum tax benefits and technological obsolescence that would normally have the automobile, computer, or machine tool replaced by a more current model. Economic conditions certainly bear on the length of time an item of capital equipment is kept in service. When the economy is depressed, equipment will be maintained much longer than if the economy is providing enough money and confidence to stimulate investment in new equipment. A depression/recession economy provides excellent opportunity for growth of the service parts business. Component reconditioning has been very profitable in a depressed economy where there is little money available to purchase new capital goods. The markets will vary, also, by industry and geography.

In Rochester, N. Y., for example, where salt is spread on the roads during the winter, car bodies begin to show rust at about three years and generally become unsafe by ten years. That will stimulate high demand for replacement of components affected by salt corrosion and reduce the need for engine components on very old cars, since they have probably been scrapped. In locations that do not have the corrosive influence of salt, cars can be driven much longer and, therefore, need replacement parts for engines and other wearout items that may still be running after many years of use. For instance, DC-3 airplanes are still popular under rugged conditions, such as in South America, 40 years after they were popular in the United States. Several organizations within the aircraft industry arrange to buy the manufacturer's and airline's inventories at the end of an aircraft's first life and then provide the necessary after-market support while the aircraft is used in other markets. The parts are normally purchased at large discounts, so that profit can be made even though turnover is low. Similar markets exist for old automobiles, computers, medical equipment, and machine tools.

Another approach to phasing out products is to reduce the menu of parts made available by stopping reorders and eliminating cosmetic and nonfunctional parts. As the machine population in the field declines, custody exposure should be reduced and parts returned to the highest level of support for subsequent central replenishment. In this phase pricing should be reviewed since demand will be reduced and smaller purchase quantities with increased carrying costs will increase the costs of supplying the parts.

End of Life

When equipment has truly reached the end of its useful life, all support items, including parts, should be terminated with the equipment. Many maintenance organizations have parts on their shelves for equipment that was disposed of years ago. A key is "where used" information on each part record. This makes it possible for a computer program to list all parts used on equipment. Make sure that the parts for discontinued equipment are not discarded if they are also used on other equipment. See Figure 6-5.

With manual record systems, the challenge is more difficult. The manufacturer's parts list, usually in the back of the equipment manual, will be one source to check. An experienced stockkeeper can go through the list and recognize whether any of those parts are in stock. Another approach, when each line item is kept on a file card, is to go through those cards about once a quarter and review the "where used" entries against the list of all equipment terminated during that period. The time interval may be longer or shorter, depending on the movement of equipment and parts during that period. If there has been more termination of equipment, then the review should be done more often.

How much should we buy? Forecast (as covered in Chapter 8) the quantity necessary to last the desired number of years. Consider populations, cancellation rates, and wearout. Cannibalization is a valid source of parts. Perhaps buying fewer parts and instead gaining flexibility by retaining tools and drawings is the best solution. Buying parts for end-of-life can add low turning units at high cost unless those end-of-life decisions are carefully made.

FAILURE MODES, EFFECTS AND CRITICALITY ANALYSIS

Changes in demand for parts generally come either from marketing factors, such as increased product population, use,

```
 2 G     R E P L A C E A B L E     P A R T S     O N     E Q U I P M E N T

ENTER EQPT(E)/PROD(P):E      #:A/C1234         DESC: AIR CONDITIONER, 5T, YORK
ENTER SCREEN VIEW(S)/PRINT COPY(P):S
ENTER ALL PARTS(A)/NOUN GROUP(N):N    DESC: SWITCH

                                                 ON
   PART #            DESCRIPTION                 HAND        SN / PO# / DWG
273056        SWITCH       ,DPDT,30A, 220V,HVY    2
273077                     ,DPST,15A,RELAY OP     3        SEE DWG 24-55AT
273999                     ,DPST,15A,AIR VENT     NO       SEE DWG 24-57AT
298811                     ,FREON PRESSURE        5
298812                     ,OVERHEAT SHUTOFF      1
298933                     ,THERMOSTAT CONTROL    4
```

Figure 6-5   Replaceable Parts Stocked for Equipment

or functions performed, or from technology-based engineering and manufacturing design deficiencies, quality, reliability, and modifications. As pointed out earlier, not all of these are due to failures of the hardware. Failure modes, effects and criticality analysis (FMECA), or FMEA as it is sometimes called (leaving out criticality), is an excellent technical tool for judging what failures will occur, what happens when they occur, and how important that is. Troubleshooting and diagnostics, service procedures, training programs, and, of course, spares-stocking forecasts can be based on this analysis. FMECA is a bottoms-up approach in that it should start with every part on the manufacturing bill of materials (BOM). The analyst will examine each one of those parts for the modes by which it may fail: for example, mechanical breakage, high voltage, corrosion, or operator damage. The effects of each of these may be different. In some cases the part may be the only failure. In other cases there may be secondary or even further failures in the chain of events. Maintenance service then has to determine how to correct the deficiency. If the corrective action involves replacing a part, then naturally service parts management is interested. The criticality calculation will include both how frequently this event is expected to occur and how important the result is (essentiality). A boiler safety valve, for example, is a very essential, safety-related, class 1 item, but it will rarely fail if it is properly tested and exercised. By the way, the valve is a good example to look at for stocking requirements. If it is a unique part, which cannot be quickly obtained, then a spare valve should be stocked ready for instant use. If, however, a telephone call can have a new valve delivered by the time the old one is removed, then it should not be stocked.

A reliability analysis and the FMECA are excellent sources of data needed to develop the contents of a spare parts kit. The kit should contain those parts necessary to provide a certain level of protection. Protection is defined as filling a given percentage (say 70 percent) of the total demands.

7

Basic Statistics, Probability, and Risk

THE OBJECTIVE OF THIS chapter is to establish a familiarity with the fundamentals of statistics as used for managing service parts. This business revolves around numbers. The customer, field engineer, and mechanic would like immediate access to 100 percent of their parts needs. To achieve this would require immense stocks of parts, people, and money; and is unreasonable in the vast majority of situations. Therefore, the numbers, representing physical parts, must be arranged, analyzed, and communicated so that humans can make necessary management decisions. We should bear in mind the statement of Sir Josiah Stamp of the English Inland Revenue Department in the early 1900s, when he said, "The government are very keen on amassing statistics. They collect them, add them, raise them to the n'th power, take the cube root, and prepare wonderful diagrams; but you must never forget that every one of these figures comes in the first instance from the village watchman, who puts down what he damn pleases." The computer era version of the same problem is "garbage in = gospel out." We must begin with good data.

ORGANIZING DATA

Humans use numbers best when they are organized in a logical sequence of individuals or groups. Assume the data shown in Table 7-1 are quantities used each week of a 13-week quarterly period for part no. 345678, which is a switch.

The first step in preparing the data for effective use is to sort the data into order from the lowest number ascending to the highest number, as shown in Table 7-2.

Range

Now look at the range of data; that is, from the smallest number to the largest. Range can be expressed in two ways:

1. As the numerical difference between the largest and the smallest; for example, $18 - 3 =$ range of 15.

2. As 3 to 18. Knowing the absolute numbers, for example, 3 through 18, is a big help in service parts management because it not only tells us what the range spread is, but it also shows the absolute values. A range of 15 between 3 and 18 is much different than the same range between 103 and 118. If the numbers are of major importance, such as for forecasting the needs for expensive parts, it may be advisable to draw a histogram that graphs the numbers.

Cumulative Distribution

For managing parts the cumulative distribution will be useful. It is constructed by adding each week's data to the previous sum, so that the data continually increase from the lowest to the highest. In this specific case, it would be termed "quarter to date data," since each week it would tell us the total used to that date. As can be seen in Table 7-3, the first week's data are the actual 9. The second week is 9 plus 3 equals 12. The third week is the previous sum of 12 plus 12 (that week's data), and so on. The slope of the cumulative distribution shows whether demand is increasing, decreasing, or stable. A uniform slope would mean steady demand and increasing rate of use would be shown by an upward ascending line. Zero demand would be shown by a horizontal line. It is, of course, not pos-

Table 7-1   Raw Data

Week	1	2	3	4	5	6	7	8	9	10	11	12	13
Use	9	3	12	15	9	7	6	18	9	8	5	6	12

Table 7-2   Ordered Data

3	5	6	6	7	8	9	9	9	12	12	15	18

sible on a cumulative distribution to have a decreasing line, unless there were returns credited to stock. A cumulative distribution plotted as percentages rather than absolute numbers will be shown as very useful for determining levels of support and what parts should be stocked to fill various percentages of demand; see Table 7-4.

Table 7-3   Cumulative Data—Actual

9	12	24	30	48	55	61	79	88	96	101	107	119

Table 7-4   Cumulative Percentages—Actual

8	9	20	33	40	46	51	66	74	81	85	90	100

The percentages are determined by dividing each number in Table 7-3 by the total of 119. For example, $9/119 = 0.0756 \cong 8$ percent. The cumulative distribtuion on ordered data is very useful for determining the probability of need for each quantity. Table 7-5 shows the cumulative percentages for the Table 7-2 ordered data. It can be seen from the data that filling 85% of the requests requires 15 units.

Table 7-5   Cumulative Percentages—Ordered

Units:	3	5	6	6	7	8	9	9	9	12	12	15	18
Unit Percent:	2	4	5	5	6	7	8	8	8	10	10	12	15
Cumulative Percent:	2	6	11	16	22	29	37	45	53	63	73	85	100

Mean, Mode, and Median

Mean is the statistical term for arithmetic average. It is calculated by

$$\bar{X} = \frac{\sum_{1}^{n} X}{N}$$

In the above example, $\bar{X} = 119/13 = 9.15 \cong 9$.

$\bar{X}$ is called "X bar" and is the symbol for mean. The "Σ" is the Greek capital letter sigma, and stands for sum or total. The small "1" on the lower right and the n at the upper right of the Σ mean to add all numbers from the first through the last. In our specific case, that would be 3 through 18, or if you were referring to the week periods, it would be from the first week through the 13th week. The X to the right of the sigma simply refers to all those actual numbers, which are termed X's. The N below the line is the number of events occurring. In our case of 13 weeks, that means 13 events and 13 numbers.

The mean tells us what the average of all the usage was. That number conveys a lot of information. Be alert, though, to the fact that a person can stand with one foot in a 120-degree oven and the other foot in a 20-degree freezer and on the average would be at a 70-degree temperature, which, in this case, would not be comfortable.

Mode is the most frequently occurring number, in this case "9." In the parts business, a frequently occurring mode probably indicates that parts are used in a pattern, such as four bearings at a time. There may be patterns within the mode, as possibly indicated by the multiples of three in our example. The median is the midpoint of the distribution. In our example of 13 numbers, look at the ordered data and count to the seventh number; that is the median, which is also 9.

Graphics

People gain information more easily from graphical displays than from lists of pure numbers. Parts analysts will find frequent use for graph paper. Computer hardware and software are rapidly becoming available and priced so that most parts analysts of the future will simply enter the data into the computer program and it will display the graphical plot on the CRT or printout. It is important that users understand the process that is being developed and be able to do it manually when a computer is not available. For the 13 numbers in our data, a column can be established for every number from either 3 (our lowest) through 18 (our highest) or from 0 through 18. For comparison purposes with other parts, the 0 through 18 is better because it shows the absolute range and where it is on the scale of all numbers. The lowest cell is 0 because it is quite possible that there will be no demand for a part during a specific period. Figure 7-1 shows the histogram of these data. This form could also be called a bar chart, since there are bars constructed for each of the number cells by filling in the blocks on the graph paper. Figure 7-2 shows a plot of the same information; it was drawn simply by connecting the midpoints of the tops of each cell. Figure 7-3 shows a plot of the cumulative distribution of the same data.

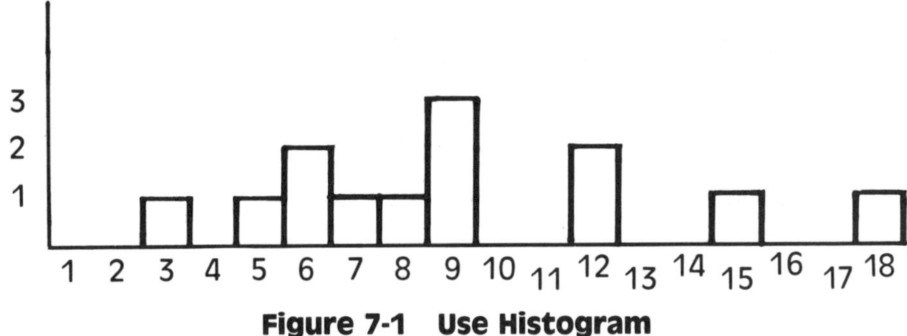

Figure 7-1   Use Histogram

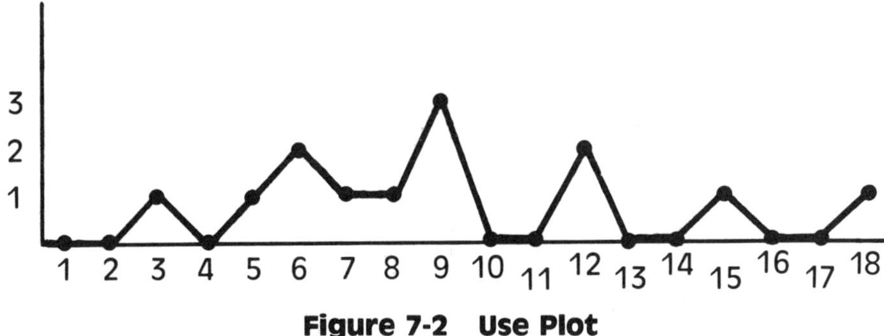

Figure 7-2   Use Plot

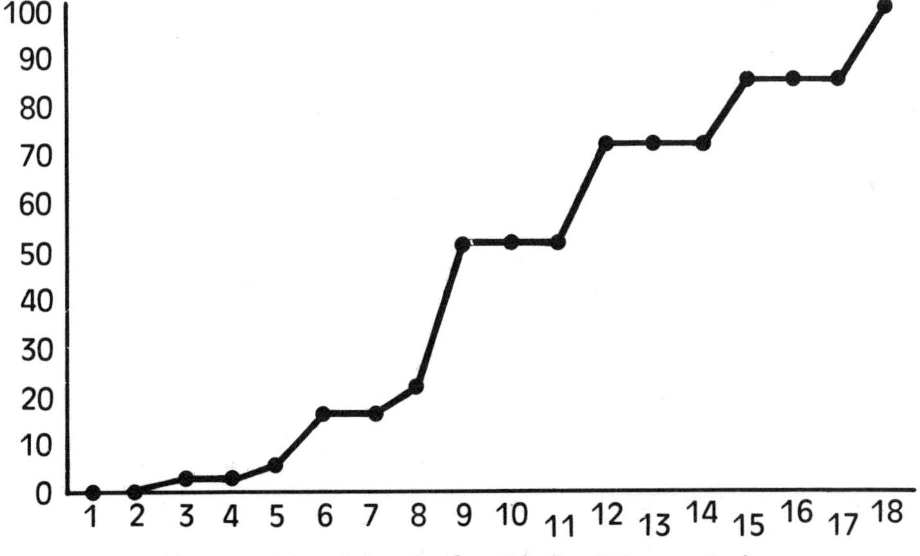

Figure 7-3   Cumulative Plot of Same Data

Parts usage will often form a normal distribution when plotted. A normal, or bell curve, distribution is evenly dispersed around the center with mean, mode, and median all being the same point, as shown in Figure 7-4.

Standard Deviation

Whether the distribution is wide and short or narrow and tall depends on the dispersion of the data from the mean. This parameter is called the "standard deviation" and is represented by S. It is also referred to as the "root-mean-square (rms) deviation," because it is the square root of the sum of the squares of the distance each number deviates from the mean. The formula is:

$$S = \frac{\sum\limits_{1}^{n}(X_j - \bar{X})^2}{N}$$

In our ordered example,

$$S = \sqrt{\frac{(3-9)^2 + (5-9)^2 + (6-9)^2 \ldots + (18-9)^2}{13}}$$

$$= \sqrt{\frac{210}{13}}$$

$$= \sqrt{16.1538}$$

$$= 4.62$$

Fortunately, many calculators and computer programs can do standard-deviation calculations very quickly. A large standard deviation indicates a wide dispersion from the mean and a wide short distribution. This also means it is harder to predict and forecast the actual values. A very small standard deviation represents closely grouped numbers and leads to increased forecasting accuracy. Standard deviations relate to probability so that in a normal distribution the area under the normal curve within one standard deviation of the mean includes 69.3 percent of all the possible values; 95.4 percent are included within two standard deviations ("two sigma") and three standard devi-

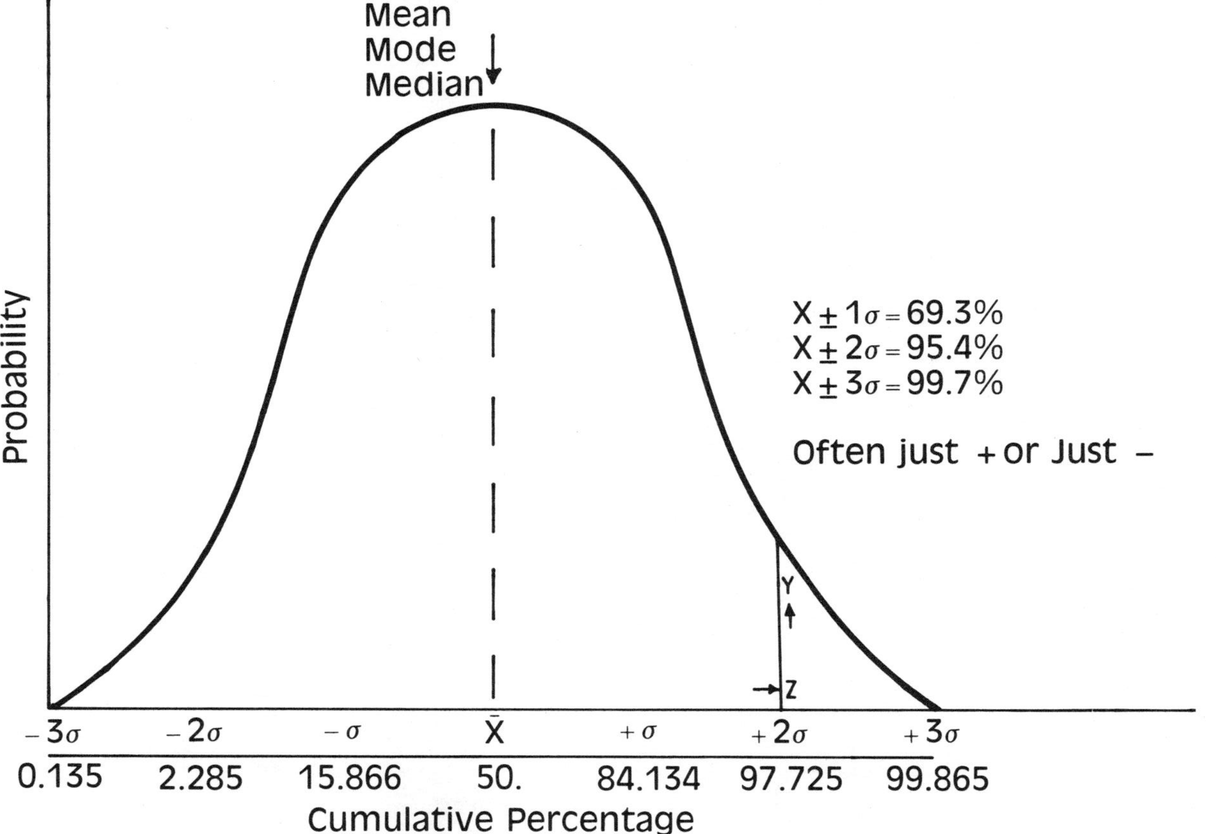

Figure 7-4   Normal Curve Areas and Ordinates Based on Deviation from the Mean

ations include 99.7 percent. Most industrial processes, such as quality control, operate at the two-standard-deviation level, and recognize that 5 percent of all the items may be outside that range. Note that, with regard to parts, we are most often interested in demands that are going to exceed the upper limit. We don't usually care if demand is very low for a short period, if there is reason to believe that it will rise again. Thus it can be termed that the 50 percent on the low side of the mean will be covered and our major interest is in the percentage on the high side. For example, if we want to cover 95 percent of the demands, then we must stock a quantity equal to the mean (9) plus two standard deviations ($2 \times 4.6 \cong 9$). In our example that would result in stocking 18, which is the highest quantity in the order. Note that it is also possible for the standard deviation to predict need for a quantity higher than actual use. The 95 percent success carries with it the 5 percent probability of failure. Five times out of 100, or one time out of 20, we should expect not to have enough on hand. Since our sample included only 13 months and not at least 20, we may not yet have seen that excessive high demand, which will probably occur later. That is where management judgment comes into making the decision based on costs and benefits.

There are, of course, other distributions, such as the log normal, which are skewed left or right, but the normal distribution is by far the most common. Those interested in other distributions should refer to a statistics textbook.

MOVING AVERAGE

Much part information is kept up to date by the use of moving averages. How many periods should be averaged? Three is a good number. More than that will dampen the effects of most recent data. The moving average is calculated by adding up to three numbers and dividing that sum by the number of events. In our example, if there is no history before the first week, then the actual is the same as the average. In the second week, we can add the totals of the first week and second week and divide by 2, so the average will be $12/2 = 6$. In the third week, we add the first three weeks and divide that sum by 3, giving a result of $24/3 = 8$. Then in each succeeding week, the newest week's total should be added and the oldest week's total should be subtracted and the result divided by 3. Weeks 2–4, for example, are $30/3 = 10$.

Table 7-6
Three-Period Moving Averages

9 6 8 10 12 10 7 10 11 12 7 6 8...

Note that moving averages (see Table 7-6) dampen the effects of change so the average never climbs or dives as fast as do the data. The standard moving average gives equal weight to all periods. If more emphasis is desired on more recent data, then weight is recommended, such as 0.6 for most recent, 0.3 for prior one, and 0.1 for prior two. Thus the data for weighted values become

9   5.4   9   12.9   11.1   ...

For example,

$$\text{Week } 3 = [(0.6 \times 12) + (0.3 \times 3) + (0.1 \times 9)]$$
$$= [(7.2) + (0.9) + (0.9)]$$
$$= 9$$

EXPONENTIAL SMOOTHING

The term "exponential smoothing" sounds very sophisticated, but in reality is simple to understand and very useful in forecasting parts. The concept is to take some portion of the difference between a result (actual) and what was expected (forecast), and apply that correction to future data. The formula is

$$\text{Forecast}_n = (\text{Actual}_{n-1} \times \alpha) + [\text{Forecast}_{n-1} \times (1 - \alpha)]$$
$$\text{let } \alpha = 0.5$$

There is considerable theoretical discussion about what the value of alpha should be. A value of 0.5 means simply that half the difference will be applied to correct the future data. An exponent of 1 means the entire correction difference should be added to the next value. The smaller the factor, the less correction will be applied; this means that more weight is given to historical data over a long term than is given to the most recent period. For example, if the actual use last period was 10, and the forecast for the same period was 14, then $(10 \times 0.5) + (14 \times 0.5) = 12$ forecast for the next period. Again, it should be remembered that the statistics are only as good as the data provided. If deviations are large, then an analyst should investigate the causes.

REGRESSION ANALYSIS

This technique is used to determine the slope of previous data changes so as to determine what was happening and help predict the future, if the same causes are expected to continue. If the points are plotted on graph paper, an eyeball estimate of a line that will connect them may be sufficient. Many pocket calculators have regression analysis programs built in and standard programs are available as computer software. Figure 7-5 shows an example of a regression line fitted between the actual points. The principle is called "least squares" because it minimizes the sum of the square roots of the distances from each actual point to the regression line.

RISK, CERTAINTY, UNCERTAINTY, AND NO KNOWLEDGE

Risk is a major consideration in data. There is risk that not all data were reported accurately. There is risk that something will change and supply or demand will vary. There is risk that all the good plans on paper will not be completely implemented. Service parts personnel must learn to manage risk.

Certainty means that we are absolutely positive about the events. Most of us can say that we are certain the sun will rise

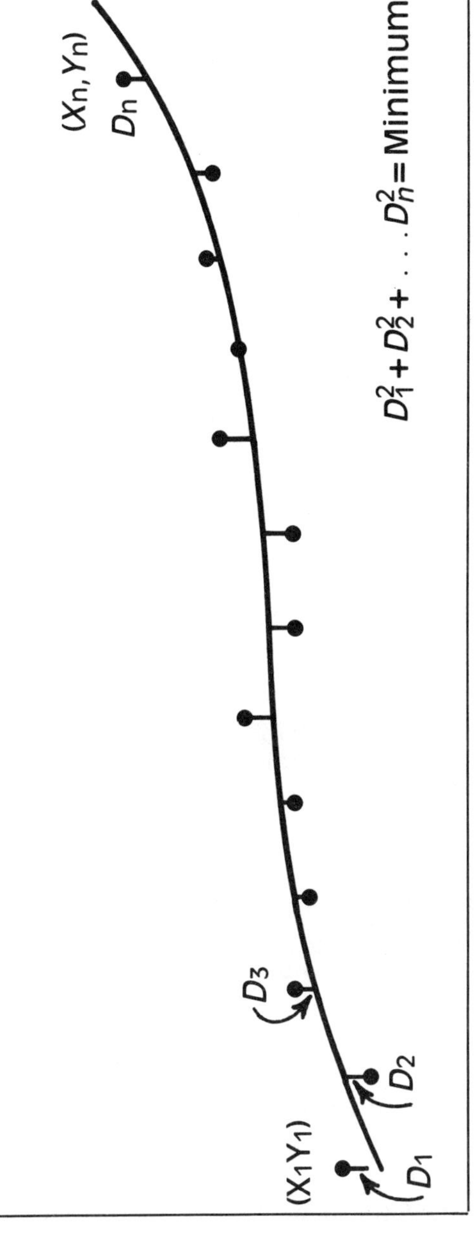

Figure 7-5   Least-Squares Regression Line

in the east tomorrow morning. We know that event has happened every day of recorded history. There is, of course, some probability that it will not happen tomorrow, but that is too small for consideration; and if it does not happen, we will probably not be in any position to worry about it. Closer to the stockroom shelves, we can say that if ten parts a week have been used for the last three years and all related causes are continuing, then the same use rate also will continue.

Anything less than certain is called "uncertain" and has a probability connected with it. A 50 percent probability means there is an equal chance of an event happening versus its not happening. This binomial (meaning two events) distribution is often referred to as the "coin toss," since flipping a coin will result in either "heads or tails," either "win or lose." Any percentage smaller than 50 percent means there is less chance of it happening; that is, it probably will not happen. Zero probability would be absolute certainty that an event cannot happen.

Most service parts considerations are uncertain. We usually deal with probabilities in the 60–95 percent range. For example, a service level of 90 percent may be desired. That means that out of 100 requests for parts, 90 of them will be filled. An illustration of probabilities used in pricing parts is shown in Figure 7-6.

Some people will say the "highest price" ($575) because they can skim off profits at that price, and then reduce the price. This might be a reasonable choice if it were possible to reduce the price without long-term loss of business. Watch out, because the second time around your customers might decide to wait until the part goes on sale, assuming they have a choice, and you could be left paying the carrying cost. If you have to set just one price and hold it, most people will select the $425 because that has a high probability of clearing out all the inventory at the best possible price. Parts managers don't like to be left holding just a few parts. They prefer to have the job completely over with.

It is important to remember that 100 percent minus percent "yes" equals percent "no." In other words, if 90 percent of parts requests are filled, then 10 percent are not filled. When selecting a percentage target, consider what the consequences of failure are. It will help to put dollar values on both the success and

What price should we bid to sell 1,000 spare parts kits:
$575, $500, or $425? Each kit costs $300.

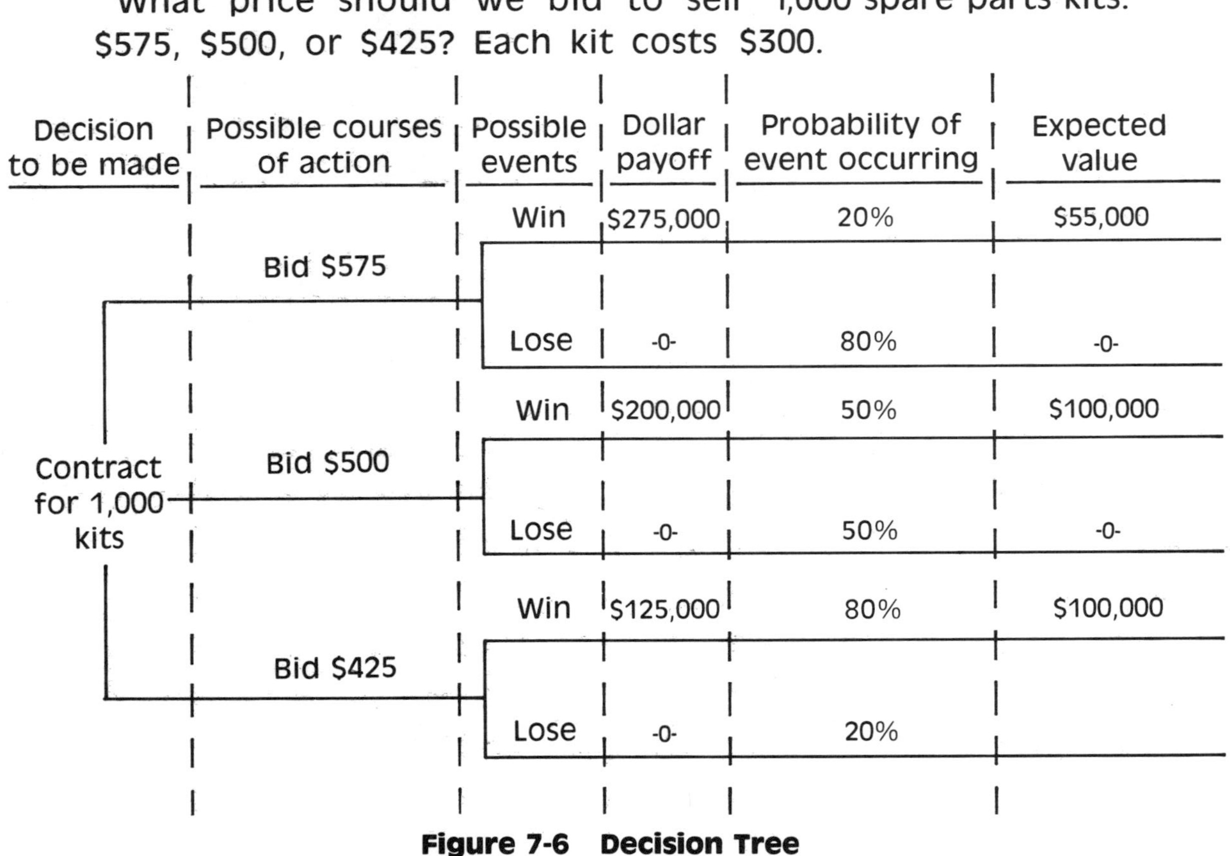

Figure 7-6   Decision Tree

failure. Too often service people become emotional over a 1 percent event and forget about the value of the other 99 percent, and probably neglect to consider the cost of achieving that final 1 percent.

MARGINAL ANALYSIS

Most service functions operate on a declining rate of return; the same idea illustrated in Pareto's principle of the critical few, the 80/20 rule, amplified in chapter 14. The greatest payoff usually comes with initial activity. Improvement per unit usually decreases with an increasing number of units. This is called "declining rate of return." In the parts business, as illustrated with the cumulative distribution, the initial improvement generally comes at relatively low cost but later improvements become increasingly more costly. Rating all parts in order according to their cost and value provides very desirable guidance to management.

SAMPLING

Sampling is a technique of utilizing a representative portion of a total population to make judgments about the rest of the items. There are textbooks written on this subject alone. For our purposes it is sufficient to know that a quantity of 30 or more is called a large sample size. Less than 30 is called a small sample size. Small sample size requires special techniques and, of course, much higher risk, since fewer events are available on which to base judgment. Most of the techniques for forecasting and doing numeric analysis are relatively easy with large quantities. Special challenges arise with the small numbers that are typical of most maintenance service operations. In most cases the service questions is, "What can we do with what we have?" The time and money cost of getting additional data can be investigated. If the potential payoff is greater than the cost, then that may be desirable. Usually, however, service management must work with what is available and recognize the associated risks.

It should be noted that statistical analysis can be used throughout a service parts system—not just for forecasting parts demand. Other examples are as follows:

1. Analysis of rate of return of failed parts
2. Analysis of time delay between failures of parts and receipt of the failed part at the central repair point
3. Analysis of part delivery deviations (delivery date versus part order date) from manufacturing
4. Analysis of material in transit (outbound, line items, dollar value, age of order)

Statistical tools should be applied to as many processes as possible so that statistical control charts are in place for each process. When a process goes out of control (deviates from an established mean plus/minus one or two standard deviations), attention can be focused on the problem. Good management means knowing what is under control as well as what problems are currently being addressed.

Forecasting

ONCE THE PROVISIONING decision has been made that decides what parts should be stocked, the next part of that same flow series is to determine quantities and patterns. The basic methods of forecasting, which range from very simple to highly complex, include:

1. Continuing trend
2. Regression
3. Factor listing
4. Cause analysis
5. Econometrics

Trend continuation assumes that the future will repeat history. As shown in Figure 8-1, a typical use chart plots actual data as a solid line, and then assumes that the trend will continue as shown by the dashed line.

Accuracy of forecasting nearly always will be better in the short term than the long term. It may be visualized as a funnel where you are looking into the narrow outlet and what you see is a refined, consolidated view of a broad expanse. When we see or think about objects at a distance, whether physical distance or time, our views are often fuzzy until we get close. Most service organizations try to do a long-range parts forecast that extends forward to the lead time of the most basic required materials. That could be two years for some electronic components. At that point we may not know what specific printed circuit boards (PCBs) will be required, but we do know that some will be required and that they will need board materials and electronics. Given that information, vendors and internal production planning can be prepared when the detailed forecast arrives. Early long lead requirements can often be in ranges, for example, a need for a quantity of 70–100. As more information becomes available and the time for actual production draws closer, the range may be narrowed. Note also that many purchasing contracts allow production plus or minus 10 percent of the ordered quantity. This is so because many processes do allow for quality fallouts. The producer will start with a quantity of perhaps 110 PCBs in order to achieve the desired 100. Should every board go through the process per-

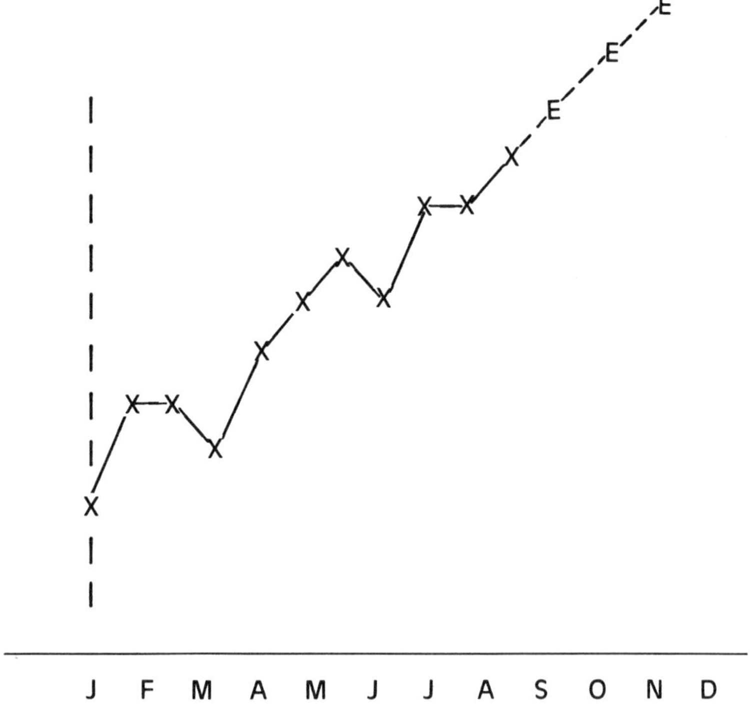

Figure 8-1 Trend Chart

fectly, the entire 110 probably would be shipped and purchased by the customer.

AVERAGE INVENTORY

The average on-hand inventory will be Order quantity/2 + Safety stock. Figure 8.2 illustrates what happens under uniform use.

The service department forecast a need for 120 timing modules during the year. Demand is essentially uniform and averages ten per month (120/12). If the part is ordered every three months as shown in Figure 8-2A, then the quantities ordered should be 30 parts at a time. Assuming a safety stock of 0, the inventory will average 30/2 = 15, as can be seen from the figure. If the same part is ordered only every six months, as shown in Figure 8-2B, then the average inventory will be 60/2 = 30, or twice as much.

CARRYING COST

The quantity kept on hand carries a cost that is termed "carrying cost" or "holding cost." Table 8-1 lists typical components of that cost.

Space considers the square feet (or cubic feet) of space occupied by the parts and support equipment. This can range anywhere from about $2 a square foot for very inexpensive warehouse space to $70–$80 a square foot in downtown New York City. Five dollars would be a good figure to use if no other figures were known. Furnishings, including shelves, files, drawer units, wall hangings, record systems, and desks, are necessary. The storeroom should be secured. This may be a steel-wire cage or a locked room with a half-door or window through which all business is transacted. The environment in which parts are stored should be clean, dry, and free of contamination. A stockkeeper should be in charge of this domain. The trade-off between paying for a stockkeeper who will take care of the parts and not having a stockkeeper means that parts

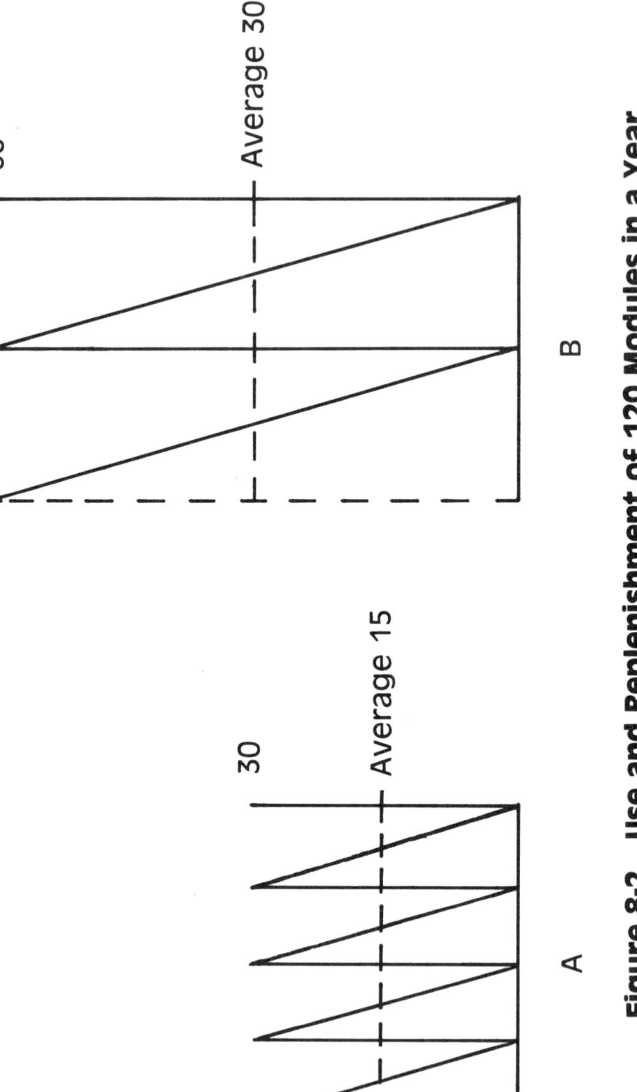

Figure 8-2 Use and Replenishment of 120 Modules in a Year

may not be available when needed, with undesirable consequences. Labor is expensive! Collecting and maintaining accurate parts data take time. If a full-time stockkeeper is employed, this person should be in charge of that function and there is little additional cost for each incremental item, up to the point where the stockkeeper is overwhelmed and additional labor must be added. The same is generally true with the information-keeping record file tubs or computer VDTs and memory involved. Increases are usually in step functions where the tub space is exceeded and a second tub must be purchased or additional computer disk capacity must be acquired. Starting and maintaining data and preparing reports involves both fixed and variable costs. In large organizations with many stock locations and catalogs, adding any line item to inventory can cost over $2000.

Insurance will be required by any prudent financial controller or lending institution to make certain that the inventory of parts can be replaced in case of disaster. Insurance on parts generally will not cover obsolescence or damage attributable to neglect or willful mishandling. If a water sprinkler were to break, any damage to parts from water spray would generally be covered. Fire and smoke, explosions of the building's heating system, earthquakes, and floods should be covered, as should burglary where forced entry can be proven. Inventory shrinkage that could be due to internal use or lack of proper security generally will not be covered. Insurance does cost money. It is acquired as compensation in case of unforeseen disasters that could wipe out a company's service business.

Table 8-1
Components of Carrying Cost

	Percent
1. Space	1.1
2. Furnishings	0.4
3. Security	0.5
4. Material handling	1.8
5. Environmental controls	0.2
6. Personnel	3.5
7. Information	1.6
8. Obsolescence	3.0
9. Damage, loss, and shrinkage	5.0
10. Insurance	1.2
11. Depreciation	1.5
12. Taxes	0.2
13. Cost of money	13.0+
	33.0

Obsolescence is decreased value due to aging such as shelf life or technical changes that render parts no longer usable. Damage occurs as a result of mishandling, banging around in car trunks, and even use of parts as diagnostic tools.

The cost of money is either the borrowing cost from a lender or the alternate value that can be gained by investing that same money in other uses. The borrowing costs can generally be figured at about 1 percent over the prime rate. That percentage usually sets the lower bound on the cost of money. Profit-oriented institutions should attempt to make as much profit on the money invested in parts inventory as they do on any other investment. Thus if a company plans new projects with a minimum return on investment of 30 percent, then the same objective should be set for parts. Adding all these costs and dividing by the parts value generally results in carrying costs of 30–40 percent. If a round number is needed without accurate data, 35 percent is a good estimate.

PROFIT VERSUS LOSS AVOIDANCE

The concept of profit readily can be understood for a whole-sale distributor or a retail dealer who cannot make a sale (profit) if the part is not available in stock when a customer comes in to buy it. Many of those customers will go elsewhere and buy an equivalent item, and possibly never return to the first dealer. Service organizations with many contract customers have committed to supply necessary parts, almost regardless of the cost. If a computer on service contract fails and needs a new memory module, the service group must obtain that module quickly even if someone at headquarters has to get out of bed on a Saturday night to go to the distribution center, take the module from stock, package it, and drive it to the airport so that it can be flown to the technical representative waiting at the customer's airport, who will rush it to the site, and install it in that computer. Those expediting costs can be many times over what normal parts distribution should cost. Forecasting then has to consider the carrying cost and the probability of needing the parts versus the profits if the parts are available or the potential loss if they are not.

ORDER COST

Surprising amounts of human energy and information flow are required each time an order is placed. Think through the following scenario.

A technician needs a part. Inventory is checked and that part is not in stock. The technician tells this to the parts clerk. Information has to be looked up in a catalog to identify the correct part number by description and manufacturer or vendor. The parts clerk fills out a purchase request. Then the purchasing function, which may be the same person or a separate organization, takes over. Several phone calls may have to be made to determine where the part can be best obtained. A purchase order must be prepared, either by hand printing, typing, or computer, and mailed to the selected supplier. The commitment must be logged for financial and physcial control. A confirming copy of the purchase order is probably required from the vendor to assure that it has been received and agreed to.

When the goods or services are received, they must be logged in by warehouse personnel. If quality control (QC) inspections are necessary, the parts must be tagged and placed in a controlled area until QC has cleared them. Then the actual parts will be placed on the warehouse shelves or transported to the requestor. Cost of ordering activity has been found to range from $18 to over $475. The high end is due primarily to extensive quality assurance requirements. There might be service organizations whose costs are under $18 a line item, but that figure is reasonable to use if no better calculation exists.

How may order costs be reduced? The best answer is through economies of scale. Consolidate orders to a supplier, so that as many similar items as possible appear on a single purchase order. If QC inspections are a major factor, then order larger quantities of those items so that they may all be inspected at once. Each inspection probably involves obtaining the applicable drawings and inspection criteria, special tools and gages, and qualified personnel. Use of blanket purchase orders is another tactic. Commit to buy all your stationery supplies for the year from a specific vendor. You should get a 10–20 percent discount, be able to telephone orders and thus expedite delivery, and receive a confirming delivery list with each individual order and a monthly invoice. Another approach, which is especially useful on common hardware, is to select the best supplier, probably on the basis of competitive bids, to keep your hardware bins stocked. This can operate in much the same way as a bread salesperson does, who comes by a grocery store once a week to check inventory and resupply stock as necessary. The salesperson will provide a list of what has been resupplied with a monthly invoice.

When doing calculations for economic order quantity and other factors that include order cost, the cost of the most common means of purchasing should be used. For example, if a part is normally ordered as one line item on a long purchase order, and is a standard item with no extra specifications or special receiving inspections, then the cost may be less than $18. If the part is uniquely sophisticated, requires extensive drawings and specifications, and must receive special quality controls, then the cost may even exceed $475. As a starting point for an average cost estimate, add the costs of all the purchasing functions involved with service parts, the stock person who requests them, warehouse receiving, quality assurance, and accounts payable, including all their overhead. Divide that total cost attributed to the acquisition of service parts by the number of line items obtained during the same period. The result will be an average order cost, which then can be tuned for the specific requirements of the groups of parts. Certainly some purchase orders contain many items and, therefore, the cost per item is lower than if only one item were ordered at a time, but the approximation is probably better than what existed previously. It is generally possible to classify parts ordering costs as low, average, or high for use where detailed costs cannot be easily obtained.

CONTROL METHODS

The methods of controlling inventory are to fix either the interval timing of when an order is placed and vary the quantity ordered, or to fix the quantity and vary the interval. The basic features of each method are shown in Table 8-2.

ECONOMIC ORDER QUANTITY (EOQ)

Figure 8-3 shows the order costs starting at 0, if no quantities are ordered (disregarding the organization that may be in place to procure parts regardless of whether or not any are ordered) and stepping upward in direct relationship to the number of orders placed during a year. One year is a convenient term to use for most of these calculations, although any period can be used, as long as it is the same for all elements.

Carrying costs are high if there are few quantities ordered during the period, because, as shown earlier, the average inventory will then be high. Carrying costs will decrease as more orders (of smaller quantities, with smaller quantities then on the shelve) are placed. The total cost curve is the sum at any order quantity of the order cost plus the carrying cost:

Total cost = Order cost + Carrying cost

It can be seen that the total cost curve starts high and decreases to a minimum point, where it starts up again. The challenge is to determine where the minimum point occurs. It should also be noted that while there is a precise minimum cost point, in practical application, due to imprecise information and the relatively small curve slope immediately on each side of the minimum point, a range around the minimum point that considers practical order quantities should be judged by a human. For example, if the calculation showed 13.5 as the quantity, an analyst would know that you can only order parts to the nearest whole number and that they are packaged in boxes of one dozen—so 12 is the adjusted EOQ that should be used.

In the case of parts that have the potential unit of measure

Table 8-2
Types of Order Models

Type of Element	Fixed Order Quantity Models	Fixed Order Interval Models
Order point (P)	Fixed	Variable
Order quantity (Q)	Fixed	Variable
Order interval (R)	Variable	Fixed

problem, a policy of ordering one order per year per part number may be the best practice. Items such as wire, liquid adhesives and cleaners, adhesive tape, and chemicals may need special packaging and labeling. Once-a-year handling may be much less disruptive to a supply point than the ordering practice suggested by the EOQ formula.

EOQ Formula

Fortunately the calculation can be performed with a relatively simple formula that is much easier than the graphical plotting. The formula is

$$\sqrt{\frac{2 \times \text{Quantity required} \times \text{Order cost}}{\text{Unit value} \times \text{Carry cost rate}}}$$

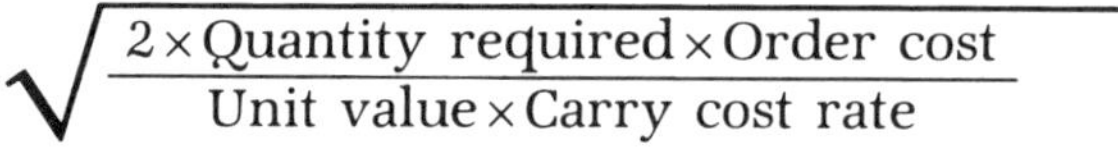

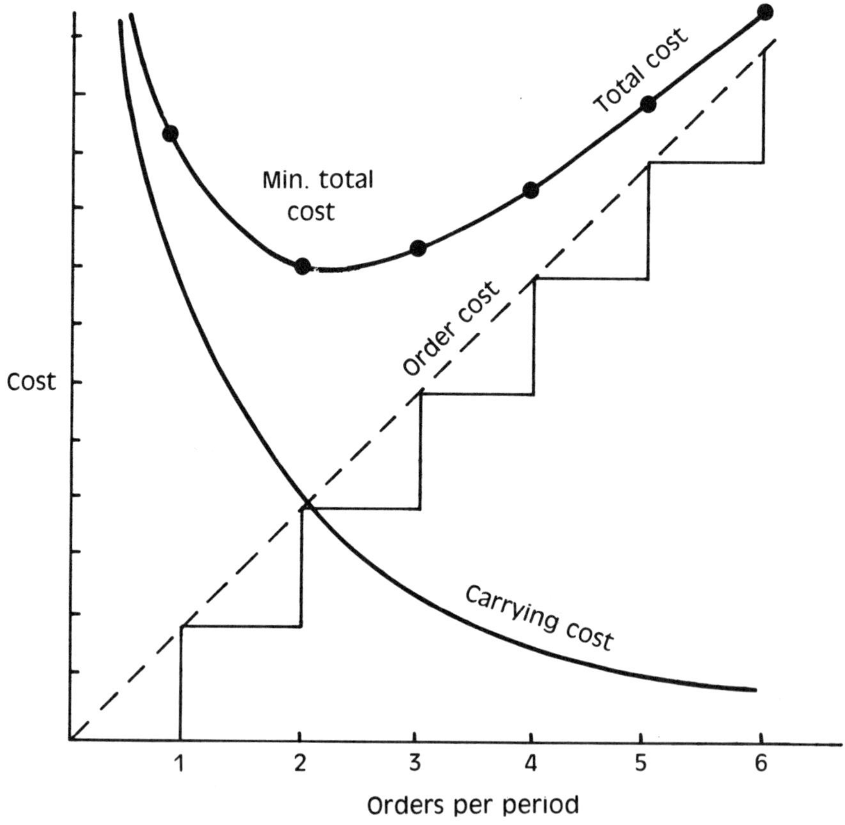

Figure 8-3 Economic Order Cost

If you enjoy mathematical derivations, you can show that this equation results from determining the point at which the slope of the total cost curve becomes zero.

Suppose we forecast a need for 50 motors during the coming year. They cost us $45 a piece, the carrying cost is 30 percent, and our purchase orders will cost $18 each. What is the best quantity to order at a time?

$$EOQ = \sqrt{(2 \times 50 \times \$18)/(\$45 \times 0.30)}$$
$$= \sqrt{1800/15}$$
$$= \sqrt{120}$$
$$\cong 11$$

Additional considerations will include obsolescence due to changes in the part that may be coming from engineering— and, of course, use changes that may be caused by sales applications in the way the product is used, special events, new technicians, production changes, and so on. Modern computer programs can calculate the EOQ as an integral part of parts management programs. From a practical viewpoint, in applying the EOQ formula, the actual order quantity placed with the supplier should not exceed one year's forecasted requirements. This policy permits several things:

1. It prevents huge orders for a part when the cost of the part is very low.
2. It prevents large bin space requirements that would be necessary if the EOQ formula were strictly enforced.
3. It reduces the chance of obsolescence of the theoretical EOQ quantity on the shelf over time.
4. It maintains an "order pattern" with the supplier by yearly ordering. It avoids long periods of time between orders and a possible vendor dropoff as a result of very infrequent orders.

In practice the EOQ should be large enough that no more than 26 deliveries are made per year (one every two weeks) to the central inventory stock point.

Persons operating a manual system or who want outside-the-computer guidance should plot representative parts and costs on graph paper as shown in Table 8-3. Use your standard carrying and order costs, possibly within parts classes, and calculate representative parts costs using those standards. Then when you need to know an EOQ, a quick glance at the table will show an approximate answer, which may be further tuned by human judgment. That figure can be plotted on standard graph paper with equal divisions so that the results show as a curve, as illustrated in Figure 8-4. Since the equation involves a square root, the same information may be plotted on log-log graph paper and will become a straight line. The divisions on the paper, of course, have been changed to the logarithmic rate. This is shown in Figure 8-5.

Choose whichever form is easier for you to understand and use for gaining accurate information. Some people prefer the equal divisions and curved line; others prefer the straight line and can interpolate well from the logarithmic grid. As an example, look at either illustration and find the EOQ for a typical part that costs $10 and is used at a rate of 100 per year. You should get an EOQ answer of 33. Then check it on the other and decide which is easier to use. Of course, a small computer is better yet!

REORDER POINT

Now that we have determined the quantity of parts to order, we need to decide when they should be ordered. This is

Table 8-3
EOQ Examples

$$EOQ = \sqrt{\frac{2 \times Order/Setup \ \$ \times Annual \ demand}{Unit \ cost \times Carry \ \%}}$$

Assume $15 order/setup and 30 percent carry costs

Unit $	Annual Demand	EOQ	Orders/ Year	Weeks/ Order	Orders, $/Yr	Carry $/Yr	Total $ Cost/Yr
1	100	100	1	52	15.00	15.00	30.00
1	1000	302	3.3	16	49.67	45.30	94.97
5	100	45	2.2	24	33.55	33.53	67.08
5	1000	141	7	7	106.07	105.75	211.82
25	10	6	1.7	31	25.00	22.50	47.50
25	100	20	5	10	75.00	75.00	150.00
25	1000	63	16	3	240.00	236.25	476.25

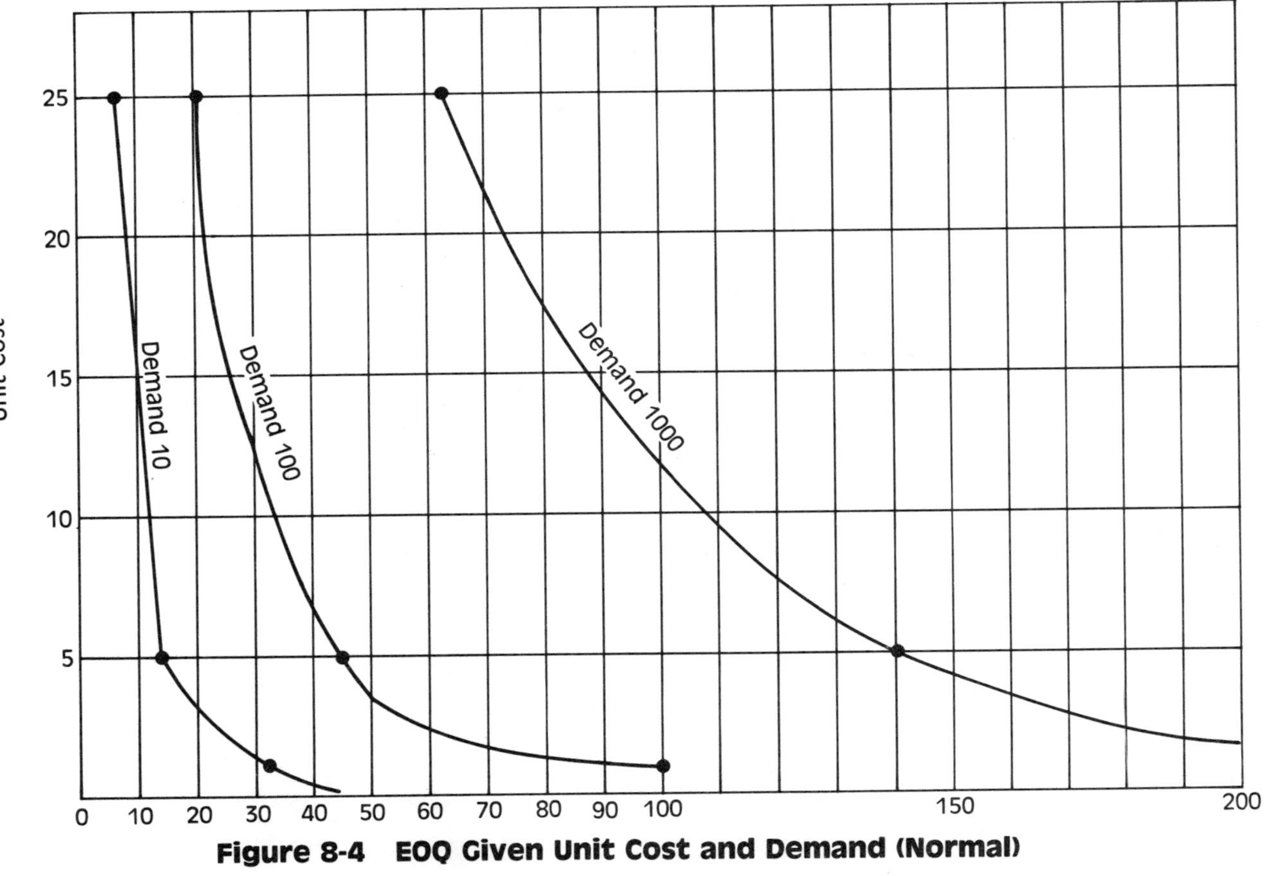

Figure 8-4   EOQ Given Unit Cost and Demand (Normal)

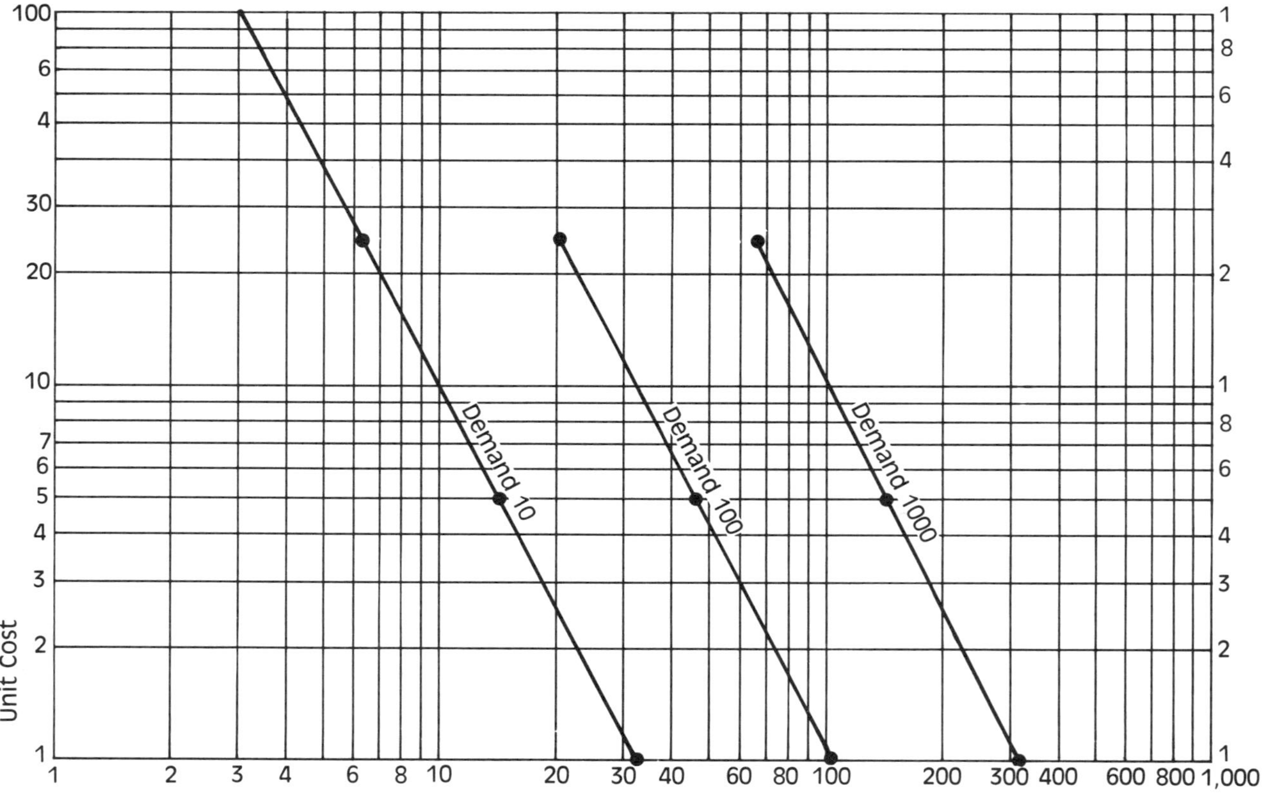

Figure 8-5 Economic Order Quantity on Log-Log Scale

called the reorder point (ROP). To calculate the reorder point, we must forecast both the quantities to be used during future time periods and the lead time. The first thing to look at is the lead time, which is composed of all the time that transpires between when the need for a part is identified and when it arrives ready for use. These intervening elements may include time to notify purchasing, get quotes, place the purchase order, transmit the purchase order to the vendor; and for the vendor to procure the item, pick and pack it, transport it to the user's location, do receiving inspection, and deliver it to the stockroom or directly to the user. The in-house time from purchase request to orders and receiving parts probably can be accurately measured. It should be less than three days for normal items, but can take up to six weeks in bureaucratic organizations where many approvals and slow paper shuffling take place. Transmitting a purchase order verbally can be done in a few hours, or by mail in, typically, three days. Once a vendor knows the purchase order is firm, a reliable estimate of delivery or production time can usually be quoted. An item off the shelf can usually be shipped out the same day if the order is received by noon. At the other extreme, quantities of nonstandard PCBs may require 18 months to produce.

The major variable is transportation. Small items may be transported by express mail or one of the several overnight package services. Larger items shipped by truck can usually arrive within a week. If parts are required from overseas, however, the transportation and the customs processing time may be several weeks. Air shipment costs, especially of heavy machine parts, can be very high. Lead time should be calculated using the most probable statistical (mode) time. Keep records on how many days there are between purchase request and parts receipt, by vendor. A computerized purchase order system can automatically gather that information by vendor and by part noun and commodity group and class. Remember though that records are generally historical data, and purchasing humans must translate that data into a forecast for an order to be placed to support the future. Lead times will normally be consistent by commodity group; thus, as a group, roller bearings will come in short supply, or motors, or electronic components. Let us assume the switches we need have a 30-day lead time (watch out for glibly stated "30 days" or "90 days" and be sure to add your internal time to what the vendor reports). During the 30 days, we forecast a typical use of 45 switches. That is an average of 1.5 per day. We realize that it is not possible to use part of a switch. What really happens is that on the first day of our month, three switches are used. On the second day none; on the third, two; on the fourth, one; and so on. To have parts continuously available, an order should be placed when the number of switches in the bin reaches 45. Figure 8-6 shows graphically what happens.

If everything holds, 30 days after the order is placed, the mechanic who receives the last part from the stockroom will turn around and bump into the UPS delivery person with the shipment of replacement parts. For many reasons, that rarely happens in the real world!

SAFETY STOCK

Since being without parts may cause expensive downtime, a form of insurance called "safety stock" will be utilized for essential parts. This means acquiring additional parts to assure that unforeseen deviations do not cause stock-out. From the normal distribution presented in Chapter 7 on statistics, it can be seen that the higher the desired customer service level is, the larger the safety stock must be to ensure that requirements are covered. As a rule of thumb, highly essential parts with lead times of over 30 days and a stock-level goal of 90 percent or above should have a two-standard-deviation safety stock. Essentiality 1 or 2 parts that have no alternative such as next higher assembly to be used if they fail, should be stocked at the two-standard-deviation level. Essentiality class 3 parts with a lead time of over 30 days should carry a safety stock of one standard deviation.

Service part demand rates are typically a combination of a few "high movers" and many "slow movers." field quantities are typically small (often one per order) and result in a demand stream similar to:

0 1 0 0 2 0 1 0 0 0 1 0 1 3 0 0 1

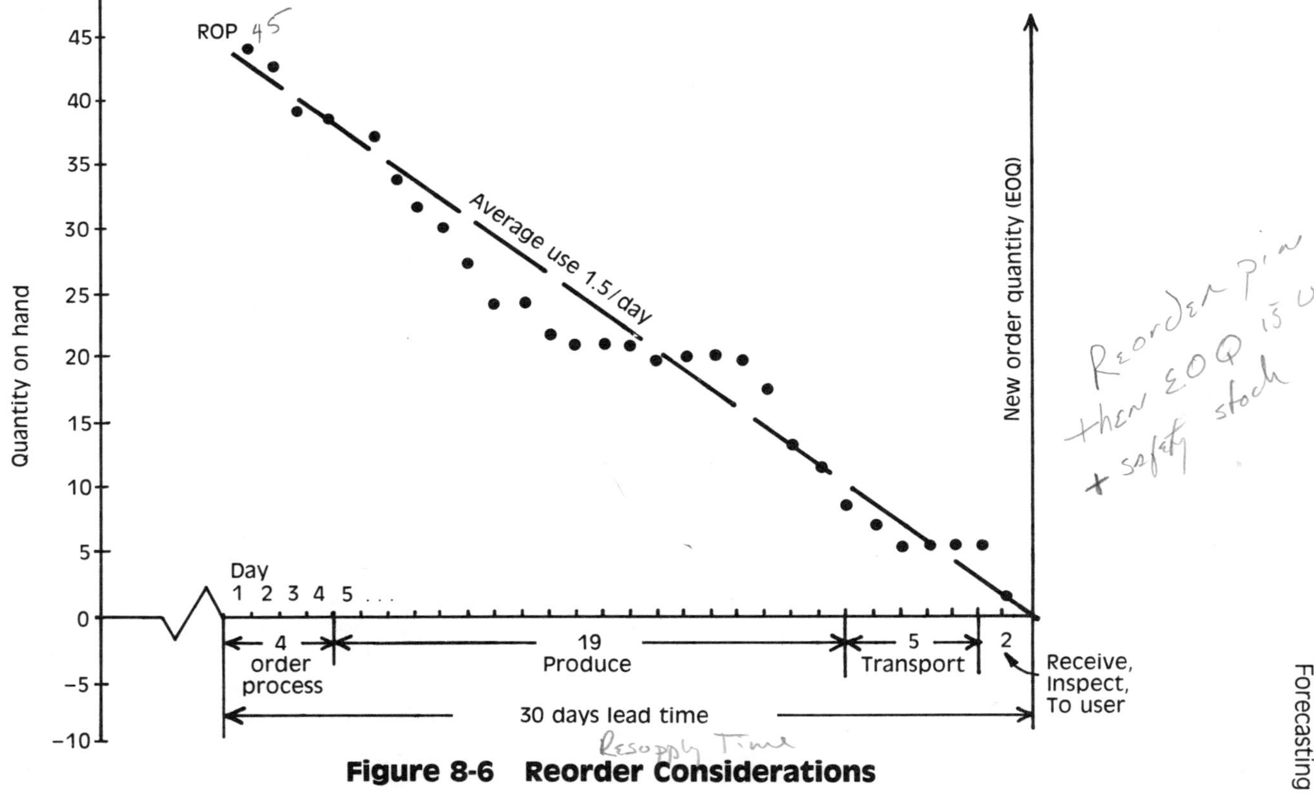

Figure 8-6   Reorder Considerations

At this point the average monthly demand is less than one and most statistical formulas for forecasting cannot be used for safety stock calculations. The quantity calculated is either zero or a fraction of a part.

There are organizations that use the concept of a minimum stock level for low-demand parts. This policy states that a minimum quantity will be held as a safety stock regardless of past demand. If demand increases over a significant period of time, the statistical safety stock will override the minimum stock quantity.

The quantity in any minimum stock level will depend on the lead time needed to obtain a replacement part if the stocked part is used. For control, only essentiality 1, 2, or 3 items should be stocked under such a policy.

Remember that those safety stock parts are going to carry the same holding cost penalties as any other part would. If they are not used, then they consume money that could be used better elsewhere. The level of safety stock is a management judgment that should be tempered by the quantitative facts of lead time, forecast, essentiality, and penalty cost if parts are not available. Let us examine some scenarios on the reorder point. Suppose, first of all, that the demand varies. Instead of averaging 1.5 switches per day, there is an increase in demand to two switches per day. That means that in the 30-day lead time the total quantity demanded will be 60 instead of 45. The parts analyst should imagine what the worst-case scenario might be and then calculate the numbers to meet that eventuality. In this case the analyst's thinking was that 60 during the period was the worst-demand case and, in fact, would cover the two standard deviations discussed earlier.

Now look at "What if?" the lead time changes. Suppose a truck strike is looming and could delay shipments by two weeks. Lead time would then be 44 days instead of 30. If demand were to continue at 1.5 units per day, then the 44 days would require 66 units. So the reorder point would be set at 66. Note that the general economic upturn also contributes to the increasing lead times as demand exceeds supply.

Observe that the deviations toward reduced demand and shorter lead time are of little importance to us. We are concerned with not stocking out of parts and are willing to accept some overage on these essential items. Now let us assume that a highly essential part is being evaluated and that everyone is pessimistic concerning lead time and demand, and the boss has stated that no essential part must be out of stock when it is needed. The worst-case conditions are thought to be a demand of two switches per day and 44 days' lead time, which multiply together to give a reorder point of 88. The difference between that quantity and the normal 45 is 43 units, nearly double the quantity. If those switches cost $9 each, the 43 safety stock switches represent $387 tied up in inventory, with a carrying cost of about $129 per year. Is that a reasonable price to pay for the safety stock insurance against stock-out? If each one of those switches costs about $2.70 a year to retain it in inventory, should the safety stock be cut to 42 units? To 41? Or to what quantity? As described in the trade-off equation, the algorithms are available to do the calculation, but management must provide the judgmental input on risk, value, and cost.

When a part quantity is reduced to the reorder point or below, the system should trigger a notice to review the situation and decide whether an order is necessary, and if it is, whether the quantity should be the EOQ or some other number. Figure 8-7 shows a typical reorder screen. Note that the order notice algorithm must consider quantity on hand plus quantity on order plus quantity in repair. This helps to avoid the problem of the organization that foolishly orders more particular parts every day without realizing that a stream of excess parts will overflow the bins. Highly embarrassing! Also, the ability to set ROP = −1 enables some nonstock parts never to show on the reorder notice.

SPECULATIVE STOCKS

Little is to be gained, and potentially a lot lost, by buying and holding service parts on the speculation that their value will increase. Several parts organizations are alleged to have purchased large quantities of electronic components containing gold and

```
              4 E     O R D E R     P A R T S  -  R O U T I N E     R O P

FOR EMPL(E)/STKRM(S): S #: S123                 NAME:  CENTRAL STOCK
                                                              QTY
        PN                    DESCRIPTION         ROP   O-H    ROQ    Y/N   NOQ
     23473        OIL         ,MOTOR,10W-40,HD      0    0     24      Y
     43266        BEARING     ,ROLLER,STNLS,3"      4    3      8      Y
     69631        TAPE        ,FRICTION            12   12     48      Y
     69632        TRANSFORMER,250W,115V,M324        1    1      4      N     1
     72007        BEARING     ,ROLLER,BRONZ,2X4     4    4     16      Y
```

Figure 8-7   Reorder Point Screen

silver in the belief that the increasing value of the precious metals would increase the parts' value. In fact, carrying costs of the excess parts (more than 30 percent per year) was greater than any increase in value. Then when the design changed and the parts became obsolete, the effort involved to redeem the precious metals was barely worth it. High-volume facilities may regain value from gold and silver components of scrap circuit boards, but the economies of scale must be great. Service parts managers should stick to managing parts and not attempt to perform financial maneuvers by speculative purchasing.

REORDER BY TIME INTERVAL

Reorder points based on time may be used instead of those based on quantity. If a distributor ships parts every Wednesday, then all parts obtained from that source should be reviewed close to the cutoff time, probably by 3:00 p.m. of the preceding day, so that the most up-to-date needs can be telephoned to the supplier. If there is confidence that the parts will be available and will be shipped on Wednesday afternoon, then the safety stocks may be kept low. Also, usually emergency support is available for extreme cases sooner than next Wednesday. The quantity ordered will be the difference between the quantity on hand and the maximum quantity that includes safety stocks. The concept of maximum/minimum stock works well for timed orders, but is not as generally effective as are ROP and EOQ.

ECONOMIC INCENTIVES

Another challenge (or opportunity, depending on your perspective) for the parts manager arises with high-volume parts when the salesperson offers to sell those $9 switches for $8.10 each in quantities of 144. Our first thought is probably that 30 switches per month divided into 144 means 4½ months' supply. The average inventory will be 72 (144/2). If the calculated

EOQ is 22 (with average inventory 11), then the difference of 61 units will have a carrying cost that must be at least offset by the discount and reduced ordering cost of the larger quantity. If the total cost equation is solved using the price of $8.10 for 144 at a time and all other factors remain the same, and the result is significantly under the total cost of 22 at a time for $9 each, then it is a good deal. In this case the calculations are

$$\text{Total cost} = \text{Acquisition cost} + \text{Carrying cost} + \text{Order cost}$$
$$= (\text{Unit cost} \times \text{Quantity}) +$$
$$(\text{Average inventory} \times \text{Carry cost/Unit}) +$$
$$\text{Order cost}$$
$$= (\text{Extended cost}) + [(\text{Quantity ordered/2}) \times$$
$$\text{Carry } \% \times \text{Unit cost}] +$$
$$[(\text{Annual use} \times \text{Order cost})/\text{Quantity per order}]$$
$$\text{Total cost}_{22} = (\$9 \times 360) + (11/2 \times 0.30 \times \$9) + (360 \times \$18)/22$$
$$= \$3240 + \$29.70 + \$294.55 = \$3564$$
$$\text{Total cost}_{144} = (8.10 \times 360) + (144/2 \times 0.30 \times \$8.10) + 360 \times \$18$$
$$= \$2916 + \$174.96 + \$45 = \$3136$$

So buy in quantities of 144 at $8.10.

Of course, you may wonder what the break-even price is. The way to answer that is to solve the algebraic equation, representing the unknown price by "P," and set the result equal to the EOQ. For example,

$$(\$P \times 360) + \left(\frac{22}{2} \times 0.30 \times \$P \right) + \frac{(360 \times \$18)}{144} = TC_{144}$$
$$(360P) + (3.3P) + (45) = \$3136$$
$$363.3P = 3091$$
$$P = \$8.51$$

MANUAL FORECASTING

Figure 8-8 shows a parts record card that can be effective in keeping use data and preparing forecasts for future need. The part in this case was a control box that was used on two different models of medical sterilizers. The top-line information gives production control data, including knowledge that the part is

PART NUMBER	PART DESCRIPTION	U I V S S U C T C T N M	FROZEN COST	AGS	SERVICE REP. STOCK QTY.	NEXT HIGHER ASSY.
588450	CONTROL BOX	P I A A S U	486.94	34	0	500390

PRODUCTS USED IN M-7 M-416
 PARTS/PRODUCT 1   1
 (x) PRESENT POP (x) 6390 (x) 8530 (x) (x) (x)
=TOTAL FIELD PARTS 14,920 (x) 80 % FIELD SUPPORTED = 11,936 TOTAL SUPPORTED PARTS

USAGE LAST YR. 233 x .6 = 140
  2ND YR. 197 x .3 = 59
  3RD YR. 162 x .1 = 16

$F_N = ($ Actual $_{N-1} \times \alpha) + \{$ Forecast $_{N-1} \times (1-\alpha)\}$

$$\alpha = .5$$

work well (handwritten)

History (handwritten)

BASE QUANTITY	1980	1981	1982	1983	1984
	215	319	322	339	356
	= # %	= # %	= # %	= # %	= # %
NET PRODUCT POP. CHANGE?	+ 180	+275	+ 120	+ 25	-50
SPECIAL PROMOTIONS?	-	-	-	-	-
PRODUCTION END?	1982	1982	1982	This Year	Ended
LONG LEAD TIME?	-	90 Days	90	90	120
PART DESIGN CHANGES ANTICIPATED?	Timer Wear	-	-	-	-
FAILURES RATE CHANGE?	2 (1.65)	Up 2%	+ 2%	+ 6%	+10%
ESSENTIALITY(1 or 2 x 1.65)					
MONTHLY PART DEMAND VARIATION?	-	-	-	-	-
					Canabalize if more needed
FORECAST	355	325	330	360	350
ACTUAL	282	318	348	351	

Figure 8-8 Parts Requirement Forecast Card

the same for both manufacturing and service. The present population is estimated on the basis of shipments of the products, and further estimates of the percentage of those that are supported with this part. Next comes the weighted average. Here more credibility is given to recent use, so a weight of 60 percent (0.6) is put on the most recent year, declining to 0.3 (30 percent) the prior year, and the remaining 0.1 (10 percent) for the third previous year. The sum of those uses gives the base quantity.

The next concern is what factors may modify the base quantity, and by how much. The marketing manager should be able to answer the questions on population, promotions, and production end. The engineering manager should be able to give advice on long lead times, design changes, failures, and essentiality. Service records will guide any seasonality or other trends. The major influence comes from the essentiality rating of 2, which carries a "rule-of-thumb" multiplier of 1.65 to give roughly a two-standard-deviation safety stock. (Statistical purists will jump up and down shaking their theory books, but it works for a starting position and is quickly refined by experience.) Multiplying the base quantity of 215 by 1.65 gives 355, without the need to include the relatively small population consideration.

Note that a forecast based on weighted history never can lead a changing use. It will always lag behind. If use is increasing, the average will respond more slowly and show fewer parts than may be needed. In this example, the weighted average of 215 is fewer than the 233 actually used last year. The safety stock makes up the difference, but it should be checked by a human to assure that projected growth or decline is not out of bounds.

The forecast quantity of 355 parts was ordered in 1980. During the year 282 parts were actually used. The challenge then is to improve the forecast with real data. The exponential smoothing equation shown in the middle right of the card in Figure 8-8 really adjusts by half of the difference between actual and forecast quantities due to an alpha factor of 0.5. That simply splits the difference in half. The numbers of our example are $(282 \times 0.5) + (355 \times 0.5) = (141) + (178) = 319$. A small modification for increasing population and failures gives a forecast of 325. Essentiality multipliers are no longer necessary. (Again, theorists may argue.) The actual use of 318 confirms that we are as close as forecasting ever deserves to get. The forecast for next year is 322 [(325 + 318)/2]. Our forecast was modified to 330 for 1982, and during that year we needed 18 more parts than predicted, for a total of 348. The fact that manufacturing uses the same configuration allows flexibility of trading a few parts back and forth, and also allows for faster scheduling in emergencies. Given the increasing use, the next year the quantity was raised to 360. As can be seen from this basic card, the process requires a mix of science and artistic judgment, with a willingness to risk both up and down.

NORMAL FLUCTUATIONS

Nearly every operation has patterns that vary with the season, temperature, weather, and planned events. These are often classed as seasonal, nonseasonal, event related, or irregular. Vacations of both user personnel and service personnel are a major factor. Fewer modifications are done during July and August because tech reps like to vacation then and so are not available to do the work. On the other hand, many factories close down for a week or two during that same period, and so equipment may be available for maintenance and parts replacement that cannot be accomplished at other times. The period from mid-December through mid-January is slow for medical products because people avoid the hospital during the holidays when possible and elective surgery is generally not scheduled then. Income-tax time has a major influence on computers, copiers, and word processors used by public accountants, lawyers, and other organizations involved in preparing tax returns. A majority of retail business is conducted during November and December so everything in those stores and warehouses must be maintained to perform well during those periods. A nuclear power-generating plant uses few parts during normal operation, but in preparation for planned shutdown will order and stock to a high point on everything from gloves to valves. The agricultural parts business reflects one of the most seasonal demand models available. It is largely dependent on water, economy, and government programs. Seeding and harvest periods may increase usage 10–50 times normal volumes. The point is that examination of the cycles that influence supported equipment can reveal when parts are likely to be required and allow forecasting to meet those needs. Figure 8-9 shows a typical computer-screen image that displays historical data and allows forecasting by month.

PLANNING AND ADJUSTING

The screen shown in Figure 8-10 presents a simple, but effective, output from a program that can calculate a future year's parts budget simply by multiplying the forecast yearly part use by the unit cost, and then summing to a grand total.

SEARCH FOR THE PERFECT FORECAST

In reviewing statistical forecasting techniques and theory, there is a tendency to become caught up in forecasting for forecasting's sake. Time, effort, and data resources are consumed in trying to locate the most precise technique. Simple techniques are often overlooked in favor of the more sophisticated.

A forecast is just that—a forecast. There are a number of other factors in any operation that tend to obscure the precision of a sophisticated forecasting technique. These detractors include:

1. Unpredictable variations in purchase order delivery actuals versus planned.
2. Inventory record errors with incorrect on-hand balances.
3. Unusual or unforeseen customer orders.
4. Buyer/analyst errors in judgment.
5. Quality problems with on-hand stock.
6. Trends not foreseen.
7. Very low demand rates.

```
                4 M     P A R T    F O R E C A S T

ENTER EMPL(E)/STKRM(S): S #:NE          NAME: NORTH EAST REGION STOCK
ID: 60301                    DESC: POWERSUPPL,FOR D125 DUPLICATOR
REV:  C
ESSN: 2      VALU CL: B     SVC ONLY: Y    SELL   $:  1,295.50    CARRY %: 35
STK LVLS:    A    B    C    D              COST   $:    647.75    CARRY $:   226.71
TGT LVL%:    98   80                       MARG   $:    647.75    ORDER $:    75.00
USE/YR:      30   8                        MARG   %:  50      RTN CRED $:   450.00
                                                               RPR COST $:   129.00

CAL ROP:     15   3                        LD DYS :  120          MEAS  : EA
ADJ ROP:          2                  E/P  EQPT USE:    HGHR ASSY:     INTERCHGBL:
                                      P    CPTR         25443          82711
CAL EOQ:     24   9                   P    DISK         25447          82711
ADJ EOQ:          12
             YR  M AV   J    F    M    A    M    J    J    A    S    O    N    D
FCST 85      32   3    2*   3    3   5*   6*   3   1*   2*   2*   1*   2*   2*
      84     33   3         3    3   5*   6*   3        2*   2*   2*   1*   2*
ACTL 84      11   3    2    2
      83     33   3    2    2    3    3    5    6    3    2    2    2    1    2
      82     34   3    2    2    3    2    6    4    5    3    1    2    2    2
COMMENTS: SEASONAL HIGH SPRING            LAST CHNG: 4/ 5/82    BY: LSB
```

Figure 8-9 Forecasting Display

```
                    A N N U A L   P A R T S   F O R E C A S T

PAGE NO. 00001
 2/20/84

PART NO              DESCRIPTION                    UNIT    FOREC   FORECAST$
                                                    COST    QTY


70001      BEARING    ,ROLLER,STAINLESS,7MM          4.75    45      213.75
70002      BALLAST    ,FLOURESCENT,INSTANT,20W       5.75    15       86.25
70003      CONNECTOR  ,ELEC,INSUL,MUL TP,COP#4       8.99    75      674.25
70004      SWITCH     ,DPST,15 AMP,110V              5.00    25      125.00
70005      SWITCH     ,SPST,30 AMP,220V              2.00    85      170.00
70007      BEARING    ,ROLLER,STAINLESS,SEALED      54.00     3      162.00
80001      FILTER     ,HEPA,CLASS100,LAM HOOD      400.00    15     6000.00
80002      FILTER     ,50 MICRON,VENT#5             34.00    12      408.00
90001      FILTER     ,OIL,1982 FORD PICKUP          7.00    18      126.00

**TOTAL**                                                            7965.25
```

Figure 8-10 Annual Parts Budget Forecast

All of these factors tend to distort the precision of the most sophisticated model and make it perform as a less sophisticated tool. A combination of models may be useful, with a particular model applied only when certain demand conditions exist. An effort to control the detractors should be made when installing a forecasting technique. Control can mean a large payback and provide the basis for a better forecast regardless of the model used.

SUMMARY

Forecasting is an art as well as a science. Information, especially that manipulated by computer, can be very useful. An experienced parts analyst, however, is the most valuable asset possible. Historical use patterns should be examined in order to list the causes of variation. Then the "What if?" question should be asked about the future.

Long-range forecasts should be prepared at least a year ahead and adjusted quarterly until the lead time for acquiring a particular part, at which point they must be turned into an order decision. Forecasting operates on information, statistics, and probabilities. Economics has a strong influence. The judgment of experienced analysts and managers is key to providing accurate guidance. Since no human or presently developed system should expect to be correct all the time, and even the "right" decision may result in shortages; there must be flexibility to get that needed part from repair, the next higher assembly, or an alternate source.

9

Physical Stocks

CHALLENGES

Deciding the best location or locations for service parts is a challenge. The extremes can range from every individual tech rep and mechanic carrying an extensive supply, to a centralized supply with rapid delivery to the user. Another question arises as to who should control the parts stocks: maintenance service, production, or a separate materials management organization. Then what physical facilities are required effectively to receive, store, control, pick, pack, and issue parts?

A stockroom to support factory maintenance or local field service should be located convenient to the people it serves and transportation because of the shipments in and out. The most important consideration is the ability to get parts to the requiring location rapidly and avoid labor delays and nonproductive downtime while waiting for parts. It is better to have one good stock center than to allow each function to have its own caché of parts. Individual stocks may begin as a well-organized project, but generally revert to a chaotic collection of materials where many items are ordered because the allegedly stocked one that is "here somewhere" cannot be found.

CONTROL

In theory, one stockroom should have parts both for production and to support service on those same products. Economies of scale would be better than with separate stocks. One safety stock, one group of people, consolidated receiving and inspection, all help keep costs low. Unfortunately consolidation rarely works. Why not? The answer is, because of distrust, lack of confidence, politics, favoritism, and poor management. A split of service to an independent stock will usually occur as the number of parts, amount of activity, and differences in parts configuration between production and service increase.

In an industrial plant, facility, or separate service organization, the parts should be controlled strictly by the maintenance organization. A durable-product manufacturer that supports its own products with after-sales service probably will establish an independent logistics/parts/support function as service revenues approach $25 million.

FACILITY LAYOUT

"Bin" is a common word denoting the location where a part is stored. It may be a shelf, a section of warehouse rack, a floor pallet, or a small container drawer. Every part should be linked to a physical location so that it can be found rapidly. There should never be more than one part in a physical location. If a shelf needs to carry several line items, then sub-bin designations should be applied. The physical layout usually can be designated by the row, vertical section within the row, and then shelf within that section. For example, A1A would be found in the first row (A), first section (1), bottom (A) shelf. The number of digits assigned to the sections depends on how many rows there are. Generally two digits, which allow up to 99 rows, are adequate. The use of alpha characters in the designation has the potential, in a single digit, of including 26 locations since there are 26 letters, A–Z, in the alphabet. Both people and computers are capable of determining the sequence whether it is 1, 2, 3, or A, B, C; although stockroom people are slightly faster on numbers than on alpha characters. There is little practical difference between a system with bin designations as alpha-numeric-alpha or numeric-numeric-numeric or numeric-alpha-numeric. All-alpha designations are more difficult to work with because the phonetic pronounciation of the letters by different people injects errors. An all-numeric sequence is preferred for rapid data entry because all the numbers can be keyed from the numeric pad on a computer terminal. This is faster than seeking alpha characters or a combination of alpha and numerics. If an all-numeric system is used, it should be stated as three different groups, for example, 3-21-13.

Start with the first designated row nearest the main entrance. The shelf units should start nearest the main aisle. Figure 9-1 shows a typical layout.

The pickers can work most effectively if the parts to be selected are arranged on the pick list in bin order. The picker on his or her physical route through the facility can pick both sides of an aisle. For example, section 1 of row B would be directly across from section 1 of row C and all the parts in section 1 of either row B or C should be picked before section 2 in any of those rows. If the rack sections are numeric, then another option is to give the aisle a designation and put even-numbered sections on one side and odd numbers on the other. When the picker completes picking all parts from rows B and C and turns the corner to come down between rows D and E, he or she is starting at the high end and working down. To make the picking easier, the parts in rows D and E could be listed in descending order. For example, if the sections are numbered, then in rows D and E, section 9 could be listed before 8, and so on. A computer is obviously helpful in doing these sorted listings. The end objective of any stockroom location and picking system should be to make the job as accurate, easy, and fast as possible.

Separate receiving and issuing traffic patterns are advisable. Most items will be received in large quantities contained in big

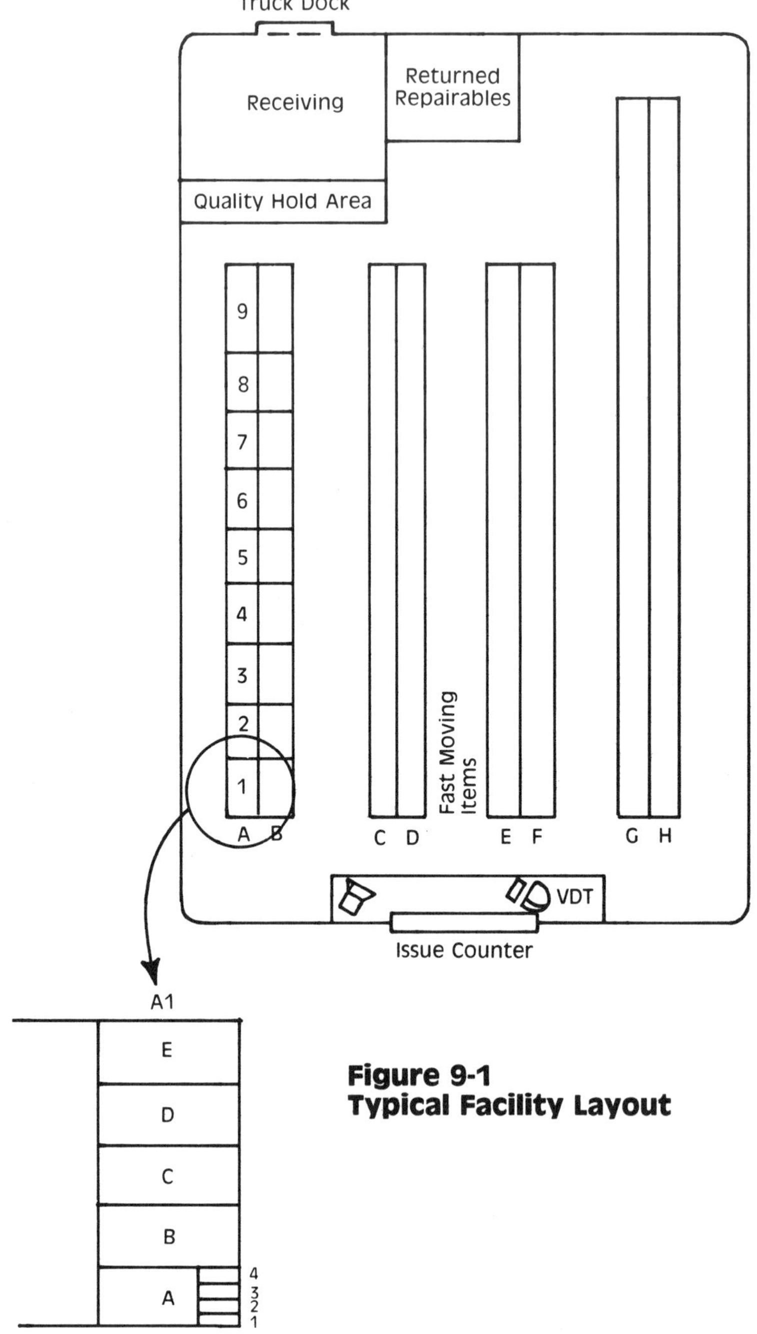

**Figure 9-1
Typical Facility Layout**

bulky containers, often on pallets, and will be issued as individual items. The receiving area should have easy access, if possible, to a loading dock so that delivery trucks can unload directly into a receiving area. Any materials that require quality control receiving inspection should be put in a separate, preferably secured, cage until they are inspected.

Since the receiving papers usually indicate only the number of packages and not their contents, a responsible person should inventory the contents against the shipping invoice to assure that what was ordered is what was received. Any discrepancies should be noted for acceptance or follow-up. It is standard procedure on large-quantity items to allow overage up to 10 percent above the quantity ordered. If 100 PCBs were ordered, the manufacturer probably started the production run with enough to assure that after quality rejects the 100 would still be achieved. If the process were working well, then perhaps an additional five boards would be produced and shipped. Quantities received over and under, damaged materials, and back orders all need to be identified and followed up.

Once parts have been checked and passed by quality assurance, they should be put on the shelves. Wherever possible they should be put in the back of the same bins that hold earlier parts. Why in the back? Because first in-first out (FIFO) is a good physical rule for any parts. That means that the older parts will be in the front of the bin where they will be easily picked and used first.

There are other methods to achieve FIFO, such as lot control, where each incoming shipment is assigned a separate bin, and outgoing orders are picked from the oldest lot. Date stamping puts labels on parts when they are received and pickers are instructed to pick the stock with the oldest date. Shelf-life-sensitive components may also be separated physically and those items put under the management of a few specially trained people who control access to them.

LOCATION ASSIGNMENT

Location of parts within a stock area should be based on four functions: safety, frequency of picking, size, and weight. Hazardous substances should be put in a safe area immediately. Parts picked most frequently should be kept near the issuing counter so the stock clerk can get them quickly. An analysis of the numbers of parts processed in the past 12 months by part numbers will assist bin location selection. High movers can be stocked in the most accessible areas, with slow movers assigned to more remote locations. Large, bulky items usually will be kept near the entrance through which they enter and leave. If they are light, they may be put high on racks where they do not use space that is better utilized by small items; otherwise they are probably kept on the floor or in a special pallet area. Heavy items should be kept near the floor so they can be handled easily and safely. Fast moving, frequently issued parts should be stored close to the place of need.

The development of a forecast for each part can be very valuable in bin location assignment. Once the EOQ and safety stock quantities have been determined, the minimum and maximum stock level forecasts are generally:

Minimum stock = Safety stock
Maximum stock = Safety stock + Reorder quantity + Reorder point

As forecasting techniques and methods improve, a week-by-week stock projection is possible. Both the minimum and maximum stock levels can be forecast for the planning period. If length–width–height measurements are developed on the database, then several plans can be developed:

1. Cube (volume), to show total storage volume required.
2. Bin size mix, to show bin sizes needed.
3. Bin change, to show when present bin size will become too small.

Packaged part volume (cubic feet) × Maximum forecast on-hand quantity = Part space requirements (maximum cubic feet)

Placing similar parts together can make sense in a small stockroom that has no computer system and is often accessed by the personnel who need to identify and select the specific part themselves. For example, all bearings might be placed in orderly sequence so that they can be compared easily and the correct one selected. In a computerized stockroom that handles a large volume of picking and issues, relational shelving can cause problems. This is so because a picker working from a computerized number list probably moves the cart rapidly down the row and literally picks the parts on the run. If the order calls for part 05555 in bin 8-22-13, it is relatively easy to miss bin 13 and pick a part out of bin 14 instead. If the number and description are significantly different, the error should be identified by the picker who got a relay instead of a bearing, or certainly by the quality checks at packing.

In most stockrooms it is easy for a human who has a part to store to find the first vacant rack and assign the part to that location. There may be a general area for standard normally stocked parts and a separate section for those special ordered direct to projects. Usually the stockkeeper looking for a place to put a new part merely starts at bin 1A and walks down the shelf area until a vacant spot is found, and then places the part in that bin location and notes the identifier for stockkeeping records. If the part is already stocked, the record card or receipt stock slip should show the bin where the new quantities are to be put.

A computer program may be developed for larger warehouses that can assign parts to the available locations. To do this the database on each part must include the length, width, and height of each part in its shelf package. The computer program then calculates the space required and searches its list of available bins to find the next large enough location, to which it assigns the parts. Naturally it does look first at any existing locations for that part to see if the shipment can be placed there, or if all the quantities of that part should be moved to another location where they can be stored together.

As picking of such orders probably also will be computerized, the pick list is organized in the most efficient sequence. The warehouse may also be automated with computer-directed narrow-aisle fork trucks that can move rapidly in both horizontal and vertical directions. Those systems are vital for high-volume, well-packaged products such as for a mail-order house where large volumes of orders must be rapidly shelved or picked. They are less justifiable in the typical factory or field service stockroom where parts come in many shapes and sizes. Systems are being developed that automatically will retrieve the container of parts and bring it to the counter where the stock-keeper may take out the necessary quantity and then push a button to send the container back to its computer-determined location.

PHYSICAL COUNTING

A typical computer program screen is shown in Figure 9-2, and the resulting printout in figure 9-3. Note that the sort order is by location to facilitate physical movement.

After the inventory is recorded, usually on paper or hand-held terminal, the results of the physical count must be entered against the perpetual count to determine variances, if any. Figure 9-4 shows helpful information.

CYCLE COUNTING

The objective of cycle counting is to confirm on a statistical basis that parts are being properly managed and that the physical stocks on the shelves equal the perpetual stocks carried electronically in the computer. Accurate cycle counting should eliminate the need for warehouse or stockroom shutdown and complete physical count. Cycle counting is done to reduce or eliminate the errors that creep in as a result of people taking or returning parts without updating the data, of theft, and so forth. Cycle counting is a far more reliable method of inventory verification than annual physical inventories.

The cycle length is calculated by

$$(1 - \text{Accuracy level desired})/(\text{Accuracy level desired} \times P \text{ of a variance})$$

The cycle length is the time or units needed to reach the target accuracy. The target accuracy level is determined by

```
INVENTORY  COUNT  LIST  &  RESULTS  ENTRY

PRINT LIST(P), ENTER RESULTS(E), EXIT(Q):  E
RANGE FROM:  70001           TO:  90001

 PART ID:  70001          DESCRIPTION: BEARING   ,ROLLER,STAINLESS,7 MM
 LOCATION : A22B1
 INVENTORY:   5              CORRECT(Y/N/Q): Y




LIST COMPLETE. PRESS ANY KEY TO CONTINUE.
STRIKE ANY KEY TO CONTINUE OR "Q" TO EXIT.
```

Figure 9-2   Physical-Inventory-Taking Screen

```
           I N V E N T O R Y    T A K I N G    L I S T

                    PART
LOCATION            NUMBER              DESCRIPTION                           QUANTITY

A22B1               70001      BEARING    ,ROLLER,STAINLESS, 7 MM             ______
A22B2               70002      BALLAST    ,FLOURESCENT,INSTANT,20W            ______
A22B3               70003      CONNECTOR  ,ELEC,INSUL,MUL TP,COP             ______
A23B2               70005      SWITCH     ,SPST,30AMP,220V                    ______
A26B1               70004      SWITCH     ,DPST,15AMP,110V                    ______
C22B1               70007      BEARING    ,ROLLER,SEALED,STNLS,SEALED         ______
CAGE 2              80001      FILTER     ,HEPA,CLASS100,LAM HOOD,WHITSON     ______
CAGE 2              80002      FILTER     ,50 MICRON, VENT #5                 ______
DO3B2               90001      FILTER     ,OIL,1982 FORD PICKUP               ______
```

Figure 9-3   Inventory-Taking List

I N V E N T O R Y   V A R I A N C E S

PART NUMBER	NOUN DESC	PHYS QTY	PERP QTY	VAR QTY	VAR Q%	PHYSICAL $	PERPETUAL $	VARIANCE $
70002	BALLAST	5	3	2	166	28.75	17.25	11.50
70003	CONNECTOR	1	2	-1	50	8.99	17.98	-8.99
70005	SWITCH	11	9	2	122	22.00	18.00	4.00
70007	BEARING	27	24	3	112	1458.00	1296.00	162.00
80002	FILTER	6	5	1	120	204.00	170.00	34.00

VARIANCE   5   ITEMS, WHICH IS   55% OF TOTAL   9 COUNTED STOCK ITEMS
NET VARIANCE $ 202   , WHICH IS   9% OF TOTAL $ 2184   COUNTED COST

Figure 9-4   Typical Variance Report

management of auditors, and is often 98 or 99 percent. Probability of a variance is the chance of a mistake occurring during the period, for example, 1/500/week. The cycles should be based on emphasizing the parts that are most important (that contribute most to the success of the facility or can cause the most harm if they are not correctly available), and to pay most attention to them. The recommended approach is to assign every stockkeeping unit to one of three classes: A, B, or C. No more than 10 percent of the items will be class A, no more than 30 percent class B, and the remainder class C.

The cycle count will be planned and controlled so that class A parts may be physically counted every 90 days, class B parts at least every 180 days, and most parts at least once every year. Exceptions are items not picked and not put away, and possibly inexpensive items that could be counted every three to five years since more frequent counting is not justified. All cycle counts should be done as an integral part of operating the stockroom and should present little impediment to ongoing operations. In fact cycle counting should provide continual benefit in confidence that the inventory is accurate. Factors that determine the class of a stockkeeping unit, and thereby the amount of attention paid to the part and frequency of cycle count, include:

1. Essentiality
2. Forecast use for year
3. Unit cost
4. Control
5. Frequency of picks and putaways

Other factors have bearing on this challenge. They include:

6. Lead time
7. Extended cost on hand

However, lead time effect can be properly managed through reorder point and safety stock. Extended cost is related to the forecast use and unit cost that are the real drivers for our need in accuracy and control. It is better to be able to control a few things well than it is to do a poor job trying to control many, so these last two factors are not considered in the rank.

Essentiality is the importance of each stockkeeping unit to the equipment it is used on and to the plant. Every SKU should already be ranked as to essentiality as:

1. Major safety or secondary failure affect
2. Production down
3. Degraded operation
4. Cosmetic or convenience

Forecast use for the next year picks one year as the convenient forecast horizon. The future is obviously what we need to manage on, so historical use data should be updated by any special needs that may cause it to change. Unit cost of the parts is generally the cost at last purchase, but may be a weighted average.

Control provides impetus to avoid theft, obsolescence, and shelf life. Items such as batteries, raingear, copper pipe fittings, and hazardous chemicals would be rated as control class 1 to provide a high level of control. Most other parts with detailed stock accounting would be the medium control level 2, and any summary stock items, such as common hardware, would be class 3, low control.

The challenge now is to consolidate the five factors into a single A or B or C class. This is a human function that must be done by intelligent people who understand where the parts are used. The matrix in Table 9-1 should help guide the decision.

Table 9-1
Class Criteria Guide

Value class	A	B	C
Essentiality	1 or 2	3	4
Forecast use for year	500+	100–499	0–99
Unit cost	$1000+	100–999	0–99
Control	High	Medium	Low
Frequency of demand	Separate algorithm		

There are obviously some combinations of these factors that cause the value class to differ from the matrix slots shown. Any one of the factors falling in the A group should put the part into the A value class. Also, a B class part should move into the higher A class if at least four of the five factors are rated B. Likewise, a part should be rated B if two or more of the five factors fall in the B category. Otherwise the parts should be ranked C. If in doubt, push the parts to the lower level because there will be too many clamoring for priority attention. It is better to give A-level attention to the critical few. The number of A-class items should be about 10 percent of the items carried.

There should be an override on activity so that if a part has been requested (i.e., a person has gone to the bin, or in the case of zero stock, at least the part was demanded) five times for A class, ten times for B class, or 20 times for C class since the last physical count, then a physical count should be taken. This is necessary because the more activity there is, the more opportunity there is for an error to occur.

Note, however, that the bigger problems will probably come from an activity such as theft or negligence that is not entered into the computer. That is covered by routine cycle counting. Another variation that should be investigated is to stimulate the cycle count at reorder point (ROP). This is so because ROP is a very sensitive time in the part's cycle. It is the lowest quantity in inventory that normally would be checked and thus makes the physical effort easy. It is also the time at which any expediting action could be stimulated if the physical count happened to be lower than the perpetual thought it was.

The best procedure for cycle counting is to have a printed list available first thing in the morning when people arrive at the stockroom. They can then take inventory in pairs with one person reading the stock numbers and a second confirming the description of those parts, counting the parts quantity, and reporting the count to the first person, who writes it on the sheet. The data would then be entered to the computer or manual system. The persons taking inventory should not be easily able to find out what the actual quantities should be. The actual quantities should never be printed on the printout. The parts for count should be displayed in bin sequence. This means there will be gaps, of course, but the physical movement of the people counting should be in a continuous pattern from one end of the stockroom to the other. To reduce error, data for quantities being keyboarded back into the system should be entered in exactly the same order that the printout sheet follows. If there are large quantities of parts to be counted, the several pages of printout can be separated and given to different people so that each group counts perhaps only one or two pages of the total.

The program needs to determine how many parts are in each class and divide that by the number of workdays that will be available in that class period. For example, in an inventory of 15,000 parts, there might be 1000 class-A parts that have to be counted within a 90-calendar-day (three-month) period. That would normally be 63 workdays. Thus $1000/63 = 16$ class-A parts to be counted each day. If there are no more than 4000 class-B parts to be counted during 126 workdays (six months), that adds 32 parts a day. The remaining 10,000 or so C class parts to be counted every year add another 40 (10,000/252) parts. The sum of those is 88 parts a day. There will be, of course, additional counts because of activity or special questions, but 100, or even 200, parts a morning, before activity really gets started, is easily accomplished. Programs need some way to determine these percentages. They could be done by humans as a one-time task in a code table, and then on the user screen a sample size 1 or 2 or 3 picked that really refers to the number of days proportioned for which the sample is being counted. Another approach is to tell the user program how many parts should be selected in class A, how many in class B, and how many in class C, and let the program also determine the counting for frequent-access parts. A third variation is to have the program first indicate the parts that are to be counted because they are at the reorder point, then the frequent-access count list, and then the balance selected from A, B, and C classes. This should all be presented as one integrated list arranged in location sequence. It would be good to indicate which parts are selected for count because they are at reorder point and which are selected because of the frequent-access criteria.

Auditors often look at the variance analyses over a period of time to determine the need for a physical inventory. If the variances were consistently (or on the aggregate) under about 3 percent in units and 5 percent in dollars, the cycle count results would be officially accepted. If the variances were more than that, then a complete physical inventory might be ordered. This would mean running a counter for date periods for the total number of line items counted, the number that were in variance, the unit quantity of those variances, and the dollar amount of the variances. The variances are symptoms that should lead to the problems. To avoid recurrence, the problem must be solved. Problems may be as diverse as people motivation, discipline, and training; or procedures, documentation, security, or management priorities. The variance data generally could be grouped by the month, or four-week periods, that are commonly used with about a 13-period total retention, with under- and over-variances listed separately so the auditors can review the data for any trends. Typical accuracy goals are shown in Table 9-2.

**Table 9-2
Physical versus
Perpetual Stock Accuracy**

Category	Goal Percentage
Items valued at $1000 or more	99
Items valued at under $1000	95
Pilferable and sensitive items	99

$$\text{Gross shrink} = \frac{\$\text{Written up} - \$\text{Written down}}{\text{Starting balance}} = 0.5\% \text{ or less}$$

MOVING TO A NEW FACILITY

Let us assume that a new facility is being established. This may involve movement to a new building, or perhaps reestablishing the service parts operations in a better location in the existing facility. The physical layout is planned to give a smooth traffic flow to shelve and pick parts with enough space between shelves for the pickers and any vehicles to move. Storage racks, shelves, bins, and drawers are acquired to handle anticipated parts volume. Every location is uniquely identified by aisle/section/shelf/drawer/box. A decision has been made as to whether parts will be placed according to stock number sequence, by commodity group, or randomly. Naturally the high-use parts are separated from low-use parts and placed near the access point. A small stock location, such as a tech rep's van, branch stockroom, or craft shop, will probably use part number within commodity group. A larger stock location should be random storage. If the locations are randomly arranged, parts may be moved in physically with little attention to where which part goes as long as it is identified and it does not have to be retrieved for service use until all parts are in and located. If specific parts are to be grouped together or placed in a particular sequence, tags should be prepared from the parts catalog and placed on every bin. Duplicate tags are helpful so that one can be placed on the existing parts and another on the destination location. A typical tag is shown in Figure 9-5.

If a computer system is used, the first set of labels could be sorted in order of the "From" location and then easily applied in sequence to those parts. The "To" location labels could be separately sorted and applied to the new bins. That assistance will allow inexperienced persons with strong backs to go to the old location, place the parts in a move container with the tag on top, and move them to the new location. There is great value in planning such a physical move well in advance and using simple systems such as move tags to assure that parts get to the

STK NO: 1234546,
DESC: RELAY, SOLID-STATE 5W

FROM: A-21-3   TO   9 - 12 - 10
Quantity: 14
MOVED BY:

Figure 9-5   Inventory Move Tag

right location with a minimum of disruption. It may also be worth the effort to have the initial quantity of the parts indicated indelibly on the tag and to have the moving person put his/her unique number, initials, or name on the tag. Again, this sign of responsibility helps assure a thorough, accurate job. If parts are subject to theft, it also helps reduce that potential for disappearance.

Self-sticking, reusable, adhesive labels are available for rapid processing through a computer printer or typewriter. Those same labels can be prepared by hand. If most of the work will be done by hand, labels using the 3M "Scotch Post-it" yellow label pads can be worth the small added expense. Any materials cost will be insignificant compared with the amount of labor and need for accuracy. When the parts are placed in their new location, the old tag that accompanied them during movement should be removed and given to data entry personnel as audit confirmation.

If the new location is not to be controlled and parts are to be placed randomly, or possibly within zones or commodity groups and classes, then any parts that are going to a special area should be tagged but the others need not be. Once all parts are moved into the new location, inventory teams must go to each bin and record the stock number, validate the part description against that stock number, count the quantity, and enter it into the record.

BAR CODE

Bar code for part identification is a technology whose time has come. A typical type 39 bar code label is shown in Figure 9-6. The labels may be prepared by relatively inexpensive printers on adhesive label stock, and then stuck to the part bin and onto packaging. This type of label is familiar to anyone who buys groceries in a modern supermarket where the checkout person passes the bar code across the reader and sees the computer-translated name and price appear on the screen. The U.S. Department of Defense now requires all suppliers to label their parts with bar code descriptions.

The bar code is typically read in a service parts situation by use of a portable wand reader/recorder. The bar code is recorded in digital format that can be later transmitted via telephone modem or tape or radio to the central computer. Obviously, the error rate is reduced and speed is increased over what is possible by writing and/or key entry of data.

CONTAINER QUANTITY

The central service parts organization should obtain, from the manufacturer or by counting, the normal quantity per container (such as 12 eggs per carton). this can help the planner place orders in multiples of the standard container quantity, if that strategy makes sense. Nonstandard container quantities must also be ordered, but standard quantities require less labor and handling, and sustain less damage.

Figure 9-6   Typical Bar Code Label Used in Military Distribution Centers

10

Inventory Valuation

ACCOUNTING GUIDELINES

THIS INFORMATION IS INtended to provide an overview of accounting practices and the Internal Revenue Code, and interpretations of law that are applicable to service parts and materials valuation. It does not represent any attempt to practice law, but rather to provide references for further investigation by your attorney and accountant. Credit goes to Robert G. Brown for much of the following research.

Valuation of inventory is generally regulated by Internal Revenue Code (IRC) Section 446, Controlling Accounting Methods. IRC Section 471 provides specific guidance. These directions require the computation of taxable income using the same method of accounting the taxpayer utilizes in keeping other records. Regulation 1.446-1(a)(2) mandates the use of accepted accounting principles applied consistently and so that the result "clearly reflects income."

The IRC provides two general methods of valuing inventories. Regulation 1.471-2(c) permits valuation at cost or at the lower of cost or market. Under Regulation 1.471-4(a), market means the "current bid price prevailing at the date of the inventory," that is, replacement cost. In the lower of cost or market, each item must be valued at cost or market and the lower value is taken as the inventory value of that article. Once a method is chosen, it must be applied to the taxpayer's entire inventory except to those goods inventoried under the last in-first out method (LIFO). Regulation 1.471-2 (d) states that a change from one valuation method to another normally can be made only upon permission from the Commissioner.

There are three exceptions to the valuation methods:

1. Under the lower of cost or market method, if there is no open market or the market is inactive, "the taxpayer must use such evidence of a fair market price at the date or dates nearest the inventory as may be available, such as specific purchases or sales by the taxpayer or others in reasonable volume and made in good faith, or compensation paid for cancellation of contracts for purchase commitments." Reg. 1.471-4(b).

2. Under the lower of cost or market method, if the goods, although not necessarily defective or obsolete, cannot be sold for the intended price, "where the taxpayer in the regular course of business has offered for sale such merchandise at prices lower than the current price. . ., the inventory may be valued at such prices as will be determined by reference to the actual sales of the taxpayer for a reasonable period before and after the date of the inventory. Prices which vary materially from the actual prices so ascertained will not be accepted as reflecting the market." Reg. 1.471-4(b).

3. Under either method, "any goods in an inventory which are unsaleable at normal prices or unusable in the normal way because of damage, imperfections, shop wear, changes of style, odd or broken lots, or other similar causes. . . should be valued at bona fide selling prices less direct cost of disposition. . .but in any case such value shall be established by an 'actual offering of goods' for sale within 30 days after the inventory date. The taxpayer has the burden of showing the goods were defective or obsolete and must keep adequate records to support his claim." Reg. 1.471-2(c).

Unless "excess" inventory is defective in some way or obsolete (and thus by definition not truly excess); has been, or is well on its way to being, scrapped; or, on the basis of objective evidence satisfactory to the Internal Revenue Service (IRS), is worth less than regular inventory of the same type; such excess inventory must be valued in the same manner as other inventory at cost or at the lower of cost or market price (defined as replacement cost).

If the goods in question are, however, completely obsolete, the court is more liberal about the 30-day offering period. For example, in Queen City Woodworks and Lumber Company versus Crooks, 7 F. Supp. 684 (1934), the taxpayer had an inventory of equipment for magnetic radios, which, in 1928, were almost completely replaced by the far more efficient dynamic radio. The taxpayer (somewhat behind the buying public, which, according to the court, recognized late in 1928 that the magnetic radios were completely obsolete and "thenceforward there were practically no demands for the magnetic radio") continued to carry the inventory on its 1928 return at its cost price of $57,155.83 instead of its actual market price of $17,046.25. The company finally caught up with the trend in 1929 and sought a refund of taxes paid. The court permitted the refund in spite of the failure to offer the goods within the 30-day period required by the regulations. This regulation does not contemplate complete obsolescence, but that there will continue to exist a demand for the goods. This liberal construction by the courts helps a taxpayer only where complete obsolescence exists and can be proved by objective evidence. Therefore, it should not be depended upon except in such factual circumstances. In situations where inventory is not completely obsolete, the 30-day limit will be in effect.

THOR DECISION

As the management at Thor discovered, valuation of excess inventory must be in the same manner as regular inventory at cost or the lower of cost or market, unless the items can be covered under one of the exceptions. In Thor Power Tool Company versus Commissioner, 99 S. Ct. 773 (1979), the taxpayer, Thor, produced hand power tools. At the same time, Thor produced large numbers of replacement parts for the power tools to avoid the costly retooling involved in manufacturing replacement parts at a later time when demand might be better estimated.

Thor valued its inventory at the lower of cost or market, and in 1964 took a general physical inventory that indicated the inventory was generally overvalued. As a result, Thor wrote off obsolete, damaged, and defective tools and, since these items were scrapped immediately, the IRS allowed this write-off. Thor also wrote down the value of replacement parts for three unsuccessful products. This write-down was also allowed because the items were immediately sold at reduced prices.

Finally Thor wrote down its excess inventory, described by the court as regular spare parts "held in excess of any reasonably foreseeable future demand," to its estimate of the items' net realizable value. This value was determined either by an aging schedule based on historical demand or a flat-percentage write-down combined with "general business experience." The values in both cases approximated scrap value. The items, although written down at once, were neither immediately scrapped nor sold at reduced prices. Rather they were retained in inventory and sold at the original prices. The Thor Company showed that, due to the specialized nature of the tools and their replacement parts, reducing prices did not increase demand. Some of the items were eventually disposed of as scrap.

Thor contended, and the court agreed, that Thor had followed generally accepted accounting principles in its inventory treatment. The court then emphasized the two-pronged nature of the test of an accounting method. It is not enough to conform to accounting practices; the method must "clearly reflect income." In fact, this second requirement is paramount and not overcome by a showing of correctness of a method for financial accounting purposes. "Any presumptive equivalency between tax and financial accounting would be unacceptable."

Thor also contended that its write-down of inventory caused the inventory to be valued correctly at market. The court disagreed, however, pointing out that market equals replacement cost under the Code. Further, Thor could not fit under any of the exceptions. There was an active market for the inventory; the items had not been offered for sale at a reduced rate because they were in "odd lots," and they had not been offered for sale within 30 days of the inventory. "Actually, Thor's excess inventory was normal and unexceptional, and was indistinguishable from and intermingled with the inventory that was not written down."

The court concluded that whenever a taxpayer departs from replacement cost for market value, "the taxpayer must substantiate its lower inventory valuation by providing evidence of actual offerings, actual sales, or actual contract cancellations. In the absence of objective evidence of this kind, a taxpayer's assertions as to the 'market value' of its inventory are not cognizable in computing its income tax."

The court recognized Thor's dilemma with its excess inventory but was not sympathetic. Thor "deliberately manufactured its 'excess' spare parts because it judged that the marginal cost of unsalable inventory would be lower than the cost of retooling machinery should demand surpass expectations. This was a rational business judgment and, not unpredictably, Thor now has inventory it believes it cannot sell. Thor, of course, is not so confident of its prediction as to be willing to scrap the 'excess' parts now; it wants to keep them on hand, just in case. This, too, is a rational judgment, but there is no reason why the Treasury should subsidize Thor's hedging of its bets."

RELATED CASES

A subsequent case, The Will-Burt Company versus Comissioner, 38 TCM 861 (1979), indicated that the mere segregation in another storage area and tagging of excess inventory (as opposed to the mingling of inventory in Thor) was not sufficient where excess items were restored to regular inventory and sold at full price when an order came in. Segregation does not equal scrapping, selling, or offering for sale at a reduced price. Also, unlike Thor, Will-Burt had written down inventory consistently for years and the taxpayer argued the significance of this consistency. The court found that mere consistency even where practices were "consistent with best accounting practices" would not prevail where the method did not reflect income.

In Altec Corporation versus Commissioner, 36 TCM 1795 (1977), the court also rejected the consistency argument, citing Thor, then on appeal to the U.S. Supreme Court. It also made clear that inventory does not become subnormal merely by reason of being excess—"while it is true that for financial reporting purposes excess inventory does not fall within the classification of subnormal goods" for the IRS.

Since the IRS is anxious to implement the Thor requirements, and since a change from incorrect write-downs of inventory to correct methods of inventory valuations ordinarily requires the consent of the Commissioner under Section 446(e), the IRS published Rev. Proc. 80-5 and Rev. Rul. 80-60. These publications required that all taxpayers with incorrect excess inventory accountings under Thor must change their methods retroactively, beginning their first taxable year ending on or after December 25, 1979; gave blanket consent to such a change without requiring individual applications for permission from the Commissioner; and stated that failure to make such a change means "the taxpayer will have filed a return not in accordance with the law." The required method of inventory valuation for taxpayers using lower of cost or market method is to value excess inventory "at replacement cost (if lower than actual cost...) unless the goods have been scrapped, or they have been sold or offered for sale (at a lower price)" within the regulations.

COSTING

Valuing the cost of parts held in inventory can follow one or more of several options, but must be consistent once a method is chosen. The options are

1. First in-first out (FIFO)
2. Last in-first out (LIFO)
3. Average
4. Specific

Why do we care? We do because we need accurate costs as a base for pricing, for charging customers, for trading off labor cost against parts costs, and for valuing our enterprise. Parts that are expensive and have unique serial numbers are usually valued at the price paid for the specific part. A FIFO inventory valuation was used by most companies until economic inflation began to increase prices rapidly. Companies realized that they were deluding themselves by valuing a part at $5 when a new replacement part would cost $10, and they could sell the one they had for $15. They also realized that since the difference between valuation cost and selling price was taxable as profit, it was desirable to pay as little tax as possible. Therefore, it made sense to value the part as high as possible. Thus most profit-oriented companies have changed their method of valuation to LIFO. This method reduces the profit margin between cost and selling price, and also more accurately represents current market value and replacement cost of the part. Note that the physical inventory need not be kept as the accounting inventory. FIFO is usually best for physical movement of parts since it assures that oldest parts are used first. This is especially true with any perishable commodities.

Quantities of parts kept in inventory are often composed of parts that were purchased at different times and costs. This presents complexities for any manual or even computer-based system. Let us assume that we purchased a batch of five parts at $10 each and later ten parts at $15 each. A FIFO system would require that the next five parts to be used be valued at $10, and the sixth part and those following at $15. This, of course, requires a counter to assure that each part is given its appropriate valuation. An average costing system would assume that the cost of the 15 parts is the total $200 and, therefore, each part costs $13.33 ($200/15). Again the mathematics get complex because when any new parts are added to the inventory, the total price of all parts must be determined and divided by the total quantity of parts to establish a new average cost. The LIFO method simply updates the cost of all parts in the inventory when new parts are received. The price of those latest parts is assumed to be the cost of all since that is the current market value. This is the easiest system for keeping records, and should be used whenever possible.

CAPITAL AND EXPENSE

Most tax-paying organizations desire to expense as many parts as possible, since that provides an immediate deduction on the tax return. The IRS, however, limits this by allowing parts to be expensed only when they are consumed in a piece of equipment and are of no further value. There is little question that consumables such as ribbons, toner, and lubricants may be expensed when they are used on equipment. Likewise, hardware, small springs, and parts that cannot be repaired are expensed. Many service organizations are flirting with trouble by expensing parts when they are issued to car stock inventory or to local stockrooms. This is often done, of course, because the organization does not have systems able to track inventory accurately at multiple locations. The benefits of being able to control inventory will have the one-time effect of regaining value in parts previously expensed, with tax implications, but the planning and control will be worth it.

High-value parts should be capitalized, with depreciation taken according to appropriate guidelines. Normally parts can be written off using the three-year IRS ADR guidelines. This requires knowing how many capital parts are in inventory at any time so they may be properly depreciated. Repairable parts create particular challenge. Let us assume that a $500 PCB fails in the field. What is the value of that failed board? If it costs $100 to repair the board, what is its value then? Generally organizations value repairable items at 40–50 percent of their new cost. Let us assume that we value the PCB at $250, and in fact give credit in that amount to our service representative and to the customer. Thus the customer might have to pay only the difference between that $250 and the full $500 cost (i.e., $250) for the replacement PCB. A defective board should be clearly identified that it is defective so it can be accounted for at a lower cost, not to mention avoiding its installation in a machine. If our repair process costs $100, the actual value of that board is $350, but it is directly comparable to a new board costing $150 more. That difference must be capitalized if all boards are to sold at the same price. Some organizations do that; others price reconditioned boards at a lower selling price based on the lower cost of $350. That can be used as a marketing strategy, and makes technical sense if there is wearout involved, which would mean a reconditioned board is likely to have a shorter life than a new board. As in most electric components, however, a reconditioned board is probably as good or better than a new one since defective components have been eliminated by use. Do remember that even though taxes must be paid on the difference between cost and sales price, it is better to keep the actual cost low and use accounting techniques to minimize the taxable profit.

11

Procurement (In-Bound)

$\mathbf{P}$ROCUREMENT, OR PUR-chasing as many people call it, is the acquisition of materials and services necessary to operate. Part of the process was discussed in Chapter 8 under the subject of "lead time." Material purchased for service parts is obtained from manufacturers or other suppliers. In many cases firm orders are loaded in a manufacturing requirements planning (MRP) system and are produced along with products in the normal production cycle. Materials managers in a service parts organization should work closely with major suppliers to share information on requirements. Automatic transfer of data for forecasts and purchase orders greatly improves performance of a MRP-driven process as shown in Figure 11-1. Figure 11-2 illustrates the full procurement process.

MAKE OR BUY

A classic decision is the choice of whether to make a part internally or to buy it from an external vendor. Economics is the most important determinant although there long-term considerations also will be given to continued employment of personnel, use of production capacity, and assurance of supply. If a choice between make or buy exists, the equation should be solved for both situations, and the least cost with best associated conditions selected. The equation is

$$TC_{\min} = C_iD + \sqrt{2C_oC_hD(1 - D/R)}$$

where:

		Make	Buy
C_i	is $ item cost	6.00	6.10
D	is demand per period	3	3
C_o	is Order or setup $ cost	50.00	8.00
C_h	is hold cost per period, $	0.004	0.004
R	is replenish items per day	18	Infinite
T	is lead time in days	12	16

Make $= \$6.00\ (3) + \sqrt{2\ (50)\ (\$0.004)\ (3)\ [1 - 3/18]} = \$18.00 + \$1 = \$19.00$

Buy $= \$6.10\ (3) + \sqrt{2\ (8)\ (0.004)\ (3)\ (1)} = \$18.30 + \$0.44 = \$18.74$

So the least financial cost is to buy.

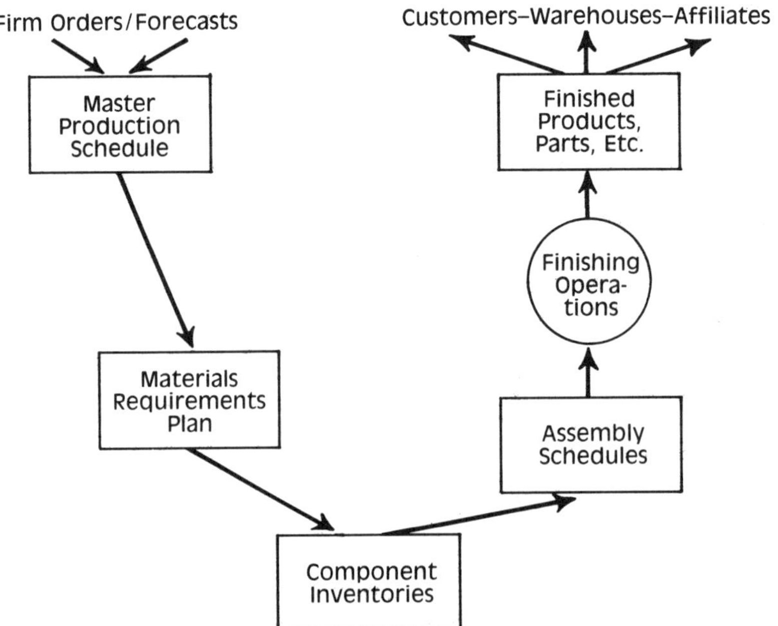

Figure 11-1   Material Flow for Service Items

Note that the rate of supply probably will vary from a fixed rate of production internally to as much as is required instantly from an outside source. Every part should be available from at least two sources if procured externally. If produced internally, then the production organization should assure backup through multiple material vendors, production lines, and so forth. The point is to avoid, if possible, having just a single source. Two or more sources lead to healthy competition that keeps prices as low as possible, and assures availability in case of strikes, transportation difficulties, material shortages, disasters, and other detrimental conditions.

Recent practices concerning vendor–user relationships indicate a trend toward using fewer vendors in an effort to improve quality and delivery. Many organizations are striving to work more closely and share knowledge of forecasts and special events. Quality will usually be the primary consideration, with delivery to schedule as number two, and price in third position.

VENDOR RECORDS

The screen image from COMMS shown in Figure 11-3 illustrates a typical vendor record. This information is linked to specific parts to provide information on all parts purchased from a vendor, and also to list vendors for a part. When requests for quotes and purchase orders are processed, the vendor information can be electronically retrieved. This same information can be kept on a Rolodex-type file or a card catalog and linked numerically to parts that are procured through each vendor. Also, some service parts operations do not become involved in the procurement process as they merely pass the purchase requests on to a separate purchasing function. Even in those situations, a recommended vendor is good information to have. If service parts management has a vendor preference, that certainly should be noted. That information on a VisiRecord® system will be kept on the traveling requisition card.

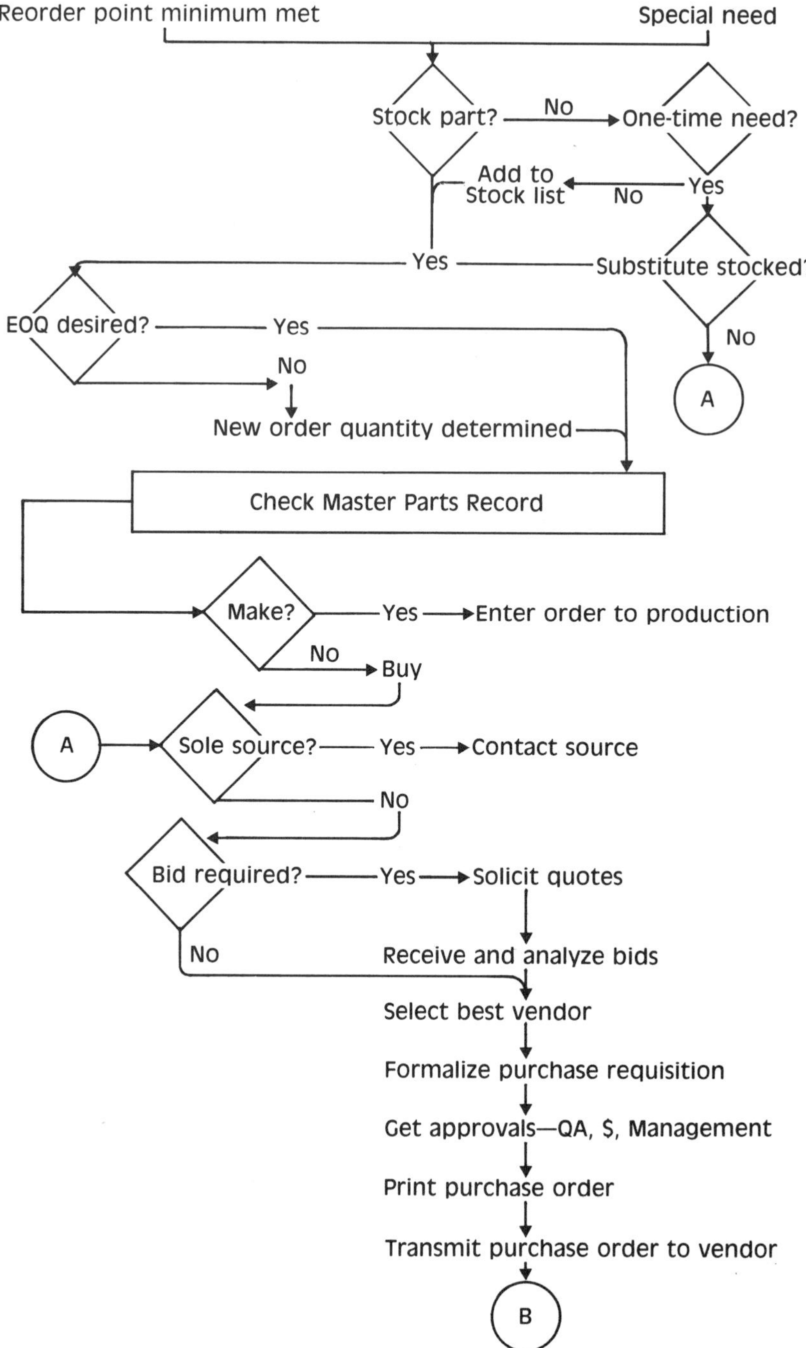

**Figure 11-2   Flow Chart of the
Service Parts Procurement Process**

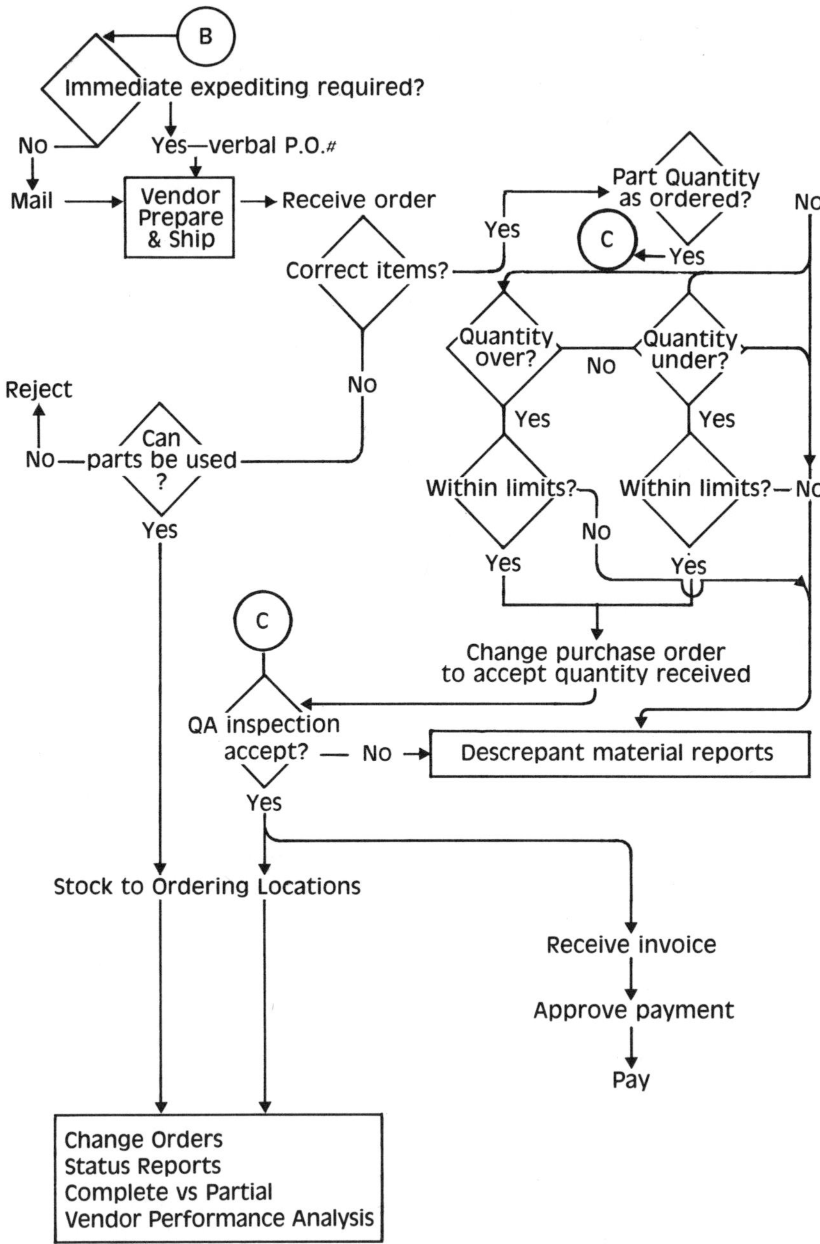

Figure 11-2   (continued)

STANDARD ORDER WORDS

The use of standard paragraphs will greatly facilitate preparation of requests for quotes, purchase requisitions, and purchase orders. A typical paragraph would be:

QA 214
All materials supplied must conform to the requirements of NRC QA regulation 214, including Amendment 1 dated 1/27/82.

Standard words can be kept in a word-processing system or a computer-based purchasing network for accurate use whenever they are required. This will reduce typing time significantly, help assure that necessary requirements are included, and keep them accurate. The standard paragraphs pertaining to specific parts should be listed on the "part record" of a computer system for automatic recall anytime they are required. Those same paragraphs would be listed on the traveling requisition cards in a paper-based system.

ORDER TERMS

Terms for buying from vendors as well as selling from your own stock generally will be net 30 days. Some vendors allow a discount such as 1/10, which means deduct 1 percent if payment is made within ten days. Note on the invoice the last date of the ten-day discount period or the 30-day net period. Most people interpret it to be 30 days from the date of invoice. Technically it should be 30 days from the date of parts delivery. Differences of opinion may arise over back-dated invoices, invoices that are not sent immediately but arrive days later and require payment by the end of 30 days (ten days from now) with penalties, and whether the date means payment must be in the vendor's accounts receivable department by that day or "the check is in the mail." The natural motivation of vendors is to try to collect payment as soon as possible, while purchasers try to pay as late as possible. Effective organizations try to sell

```
        D I S P L A Y / C H A N G E    S U P P L I E R

MFGR(M)/VNDR(V):  V            ACCT #:  V 234
                               NAME:  RELIANCE ELECTRONICS
LAST ACT :  10/13/83
      PO #:  91720             ADDR:  2700 COMMERCE DRIVE

VOL/YR $:  17,340              CITY:  LEXINGTON  STATE:  MA  ZIP: 24455
   MAX $:  25,000
                               ATTN:  BUZZ SAWYER
RANK:  1                       PHONE:  (614) 243-2555    EXT: 4210

BLANKET        EXPIRE            PAY:  ACCTS RECEIVABLE
  ORDR         DATE      ORGN. #       RELIANCE ELECTRONICS
                               ADDR:  2700 COMMERCE DRIVE
C 2965        12/31/83    123
                               CITY:  LEXINGTON STATE: MA  ZIP:  24455

                               TERMS:  DISCOUNT:  2% 15 DAYS, NET 30 DAYS
                                       BUYBACK (Y/N)?: Y      % CHARGE 15
                                       MINIMUM $: 50
                               SHIP VIA:  PICKUP  , UPS , MAIL
                                          OTHER  FEDERAL EXPRESS
```

Figure 11-3   Typical Vendor Record

the parts and receive revenues before the invoice must be paid. Cash flow is very important in service parts management.

Returns for credit sometimes will be allowed by suppliers. Electrical components are rarely accepted back again because the risk of damage and of restocking a defective component is great. Mechanical items, motors and devices that can be tested, and items in the original sealed package will often be accepted for a 10 percent restocking charge.

Most vendors also have penalty clauses for accounts that are not paid after a defined time period, usually 30 days. That interest charge is related to the cost of borrowing money. It will rise when interest rates rise, and fall when the interest rates fall. A typical rate would be 1.5 percent per month, which translates to over 18 percent a year. That means a parts organization must assure that received goods are quickly verified and the receiving documents and the invoice, should it come to them, are passed quickly to the people who pay the vendors. Most vendors give better service to customers who pay within the prescribed terms.

CHECKS AND BALANCES

Inaccuracies and theft are problems in the inventory business. A procurement system should plan to prevent such problems by checks and balances built into the process. One good rule is to have different persons involved. One person should order the part, another should receive it, and a third should pay the bills.

Dollar approval limits for purchasing should be instituted. For example, the stockroom supervisor would be able to approve purchases up to $500. Anything above that would go to the service parts manager, who would have a limit up to $5000. And orders above that dollar limit would be approved by the plant superintendent, director of logistics, or whoever is the superior manager with financial responsibility. Those limits, of course, will differ depending on the typical kinds of parts and materials that must be purchased. If the business involves $1000 PCBs that have little resale value, then the control limits may be higher. If many items are involved that have high use outside the business, such as quarts of oil and batteries, then more careful checks will be put on lower limits. It is possible to have separate guidelines that provide tighter controls for consumables than for the dedicated-use parts.

A person's signature, initials, or identifying number should be placed on every order, receipt, and issue. This helps personalize the process and assure that accurate transaction records exist. Managers must occasionally check every step of the process. A person's temptation to steal will be greatly hindered by the knowledge that someone will be checking. If a manager does not want to say that the checking is to avoid theft, it can be done under the pretense that the manager wishes to learn more about how the system operates. Most personnel are pleased that management takes enough interest in their jobs to find out, in detail, what they are doing and to ask for their opinions. A few minutes spent on the receiving dock to observe the relationship between incoming delivery personnel and receiving clerks, with attention to how carefully the receiving orders are verified, will pay dividends. Watching the detailed check of received parts may offer suggestions as to how vendors can improve the packaging of the shipments and better ways to handle internal processes. Following those parts through the entire process of locating what bin they should go to, seeing how they get there, and verifying the count in the bin help a manager understand the workings of a service parts system.

The secured stockroom previously emphasized is necessary for assuring control once parts are in the facility. The secured area may include items such as tools that are subject to pilferage, while other items, such as common hardware, can be less controlled.

VENDOR PERFORMANCE ANALYSIS

The main measures of vendor performance are delivery time, quality, and price. The priority order of those three considerations may vary among organizations. Generally, in the service parts business, quality and delivery to a reasonable schedule are more important than price. The purchasing function personnel should keep track of performance, by vendor, as measured by (1) percent of deliveries on schedule, (2) percent of shipment complete, (3) percent of received items acceptable, (4) average lead time, and (5) relative price.

Sophisticated purchasing functions have developed weighted-measurement systems that can quantitatively recommend the best vendor for a specific commodity or part. For service management purposes, our intent is to point out the major considerations. The typical person ordering service parts is either restricted as to possible sources, or must rely upon quick experienced judgment and a relationship that has been established with suppliers. Business should be spread over at least two vendors. Where possible, a dollar ceiling may be established on a particular vendor to assure that excessive reliance is not placed on a single source.

It should be remembered that while procurement must be an honest arms-length transaction process, it is always better to work with the vendors instead of against them. What is good for the service parts function should also be good for the vendor, and vice versa, so that everybody wins.

Pareto's principle of the critical few applies to vendors just as to other service concerns. Determine who the most important vendors are, work with them, and develop a mutually successful relationship that will result in quality parts delivered on time at a reasonable cost.

STATUS CODES

Requests for parts often go through many stages. These may include:

Initial request
Immediate superior
Stock control
Budget control
Quality assurance
Purchasing
Facility manager
Transmittal to vendor

Acknowledgment by vendor
In transport
In receiving
In QC
Availability for use

Time from start to finish of this process can be very short, but is often several weeks. In many organizations the lead time consists mostly of the time necessary to get a purchase order out the door to the vendor. One way to track the progress of requests is to use status codes. A simple way is to use a countdown series. If 12 events are to take place, then start with 12 when the part is first requested. Change the code to 11 when the superior has signed it, to 10 when stock control has approved it, and so on. In an order-tracking system, every part request can have a status code that then progressively counts down to code 0 (available for use), at which time it can be dropped. As procurement systems become computerized, each organization responsible for part of the function can enter its status code and see instant listing of all parts in the queue for its action. When it clears the action, the status code proceeds to the next required function. A date can be associated with each action so productivity and responsiveness of these specific functions can be evaluated. Some organizations use just a date for each step in the process. In a paper-based system, the same thing can be accomplished on a sheet of paper by listing the order and then the series of events that must occur and entering the completion date under each of those events. On a card-based system, the same may be written in, or an edge of the card may be marked to indicate each of the statuses and a colored paper clip moved from one to the next as status progresses. Obviously a computer-based system has advantages where many parts are involved or there is physical distance between facilities. It should be possible to give a telephone caller the status of any order within 3 minutes.

PARTS FOR SPECIFIC JOBS

Special order (SO) or direct to order (DTO) parts should be added to stock only if they are highly essential, are frequently used, have a long lead time, and/or are low cost. Of course, part data may be kept on the Master Parts Record so that details are readily available in case they are needed, but carried at a stock quantity of zero, meaning that it is not normally on the shelf. Occasions will frequently arise when parts and materials have to be obtained on special order.

Purchase requests and orders for these special parts normally should be tied to a specific work order, facility, or piece of equipment. In fact, work orders will often be on hold, "waiting for parts." This link to work order helps to identify all parts ordered for a project, and their status. It also facilitates expediting of the parts to the user as quickly as possible after they arrive. Accounting is also aided since cost can then be directly obligated against the work to show future commitments and then accounted for when the parts and invoice arrive.

The question will arise as to "what stock number to give these special parts." One option is to assign them a part number in the regular part number sequence structure. This option supports the shipment of the items to an automated field parts control system, control to the field, and billing from the field to the customer. Another option is to identify them with the purchase order and line item number. The latter has some advantages if the requestor will be tracking the part by the way it was ordered, which would be on the confirming copy of the purchase order. However, it is awkward if a decision is made to make those parts regular stock in the future because the number then should be assigned in the regular sequence. Either way it is helpful to have a flag on the part record that indicates this was a special order part. Some card and computer systems delete non-stock part information from the record when the quantities on hand and on order reach zero.

It is helpful to treat special parts as much like regular stock parts as possible, so that the record-keeping system will not have too many variations. It is also helpful to have a method of reviewing all special orders every six or 12 months to determine whether some of them are being ordered frequently enough to put them in regular stock. Patterns may also be observed that indicate the opportunity for better stock procedures. For example, a facility's construction and maintenance organization frequently ordered paint in special colors. This naturally led to leftovers, difficulties in matching, and other expensive problems. The opportunity was seized to standardize on a tint base that was purchased in bulk for stock and then the specific colors were mixed as required, in the amounts required. Over the long term, additional effort was being applied to standardize the colors so that only one tint of each major color would be used in the future. Similar standardization can often be applied to hardware, V-belts, gaskets, and other will-fit common items.

RECEIVING PROCEDURES

Standard operating procedures should be written, taught, and enforced to assure that the right items are received in good condition, accounted for, and accurately recorded as efficiently as possible. Most parts are delivered by truck and the driver has little or no relationship to the vendor. The driver's only mission is to hand over a carton that has the name of the receiving organization on it, and receive a signature as verification that the carton was delivered. A receiving person should at least look at the container to see that it is not obviously damaged. If it is, that fact and a description of obvious damage should be written on the way bill. As the next step, the container should be opened and the contents verified against the shipping list. For example, if the contents are listed as three each, part #1256 switch, DPDT, and two each, part #7211 switch, SPDT, then at least someone who knows that DPDT means double pole, double throw and SPDT means single pole, double throw, and how to tell the difference, should check the received parts. This will often be a quality control function staffed by inspectors with the necessary drawings, specifications, and measurement devices. Check for obvious damage and look carefully at any parts that may be damaged even though the outside container is not. Large air filters, for example, are sometimes received damaged in undamaged cartons. Any discrepant parts should be immediately tagged. Responsible persons should be notified so that a decision may be made as to use or rejection of the part. The person who made purchasing arrangements should be involved, since the vendor must be notified that the part was discrepant and, if possible, advised as to what the deficiency was and how it might be avoided in the future. If the receiving service organization incurred additional costs for rework of the part before use or other situations arise from the discrepancy, adjustment in the price should be negotiated.

The quantity received, as discussed earlier, may be over generally by 10 percent. If it is over more than that, or is under the ordered quantity, procurement management should be notified to take whatever action is necessary. In many supply organizations, the in-bound quantity must be exactly the quantity on the purchase order. If an overshipment occurs, the excess material is set aside for a disposition initiated by the buyer, such as return, hold, or receive.

Early shipments (over two weeks prior to requested delivery date) should be held for disposition by the buyer. An early shipment, unless specifically authorized, will tend to disrupt the normal flow of in-bound material. It can have a negative impact on prepackaging activity, bin assignment, bin sizing, and other warehousing. In addition, early shipments cause unnecessary carrying of inventory. Dispositions for excessively early deliveries can include return, hold until requested delivery date, and receive if needed. The lever normally used to obtain just-in-time delivery is to hold the supplier's invoice and delay payment until the specified delivery date and terms.

INTERNATIONAL DEALINGS

Many products and parts today are procured in the world marketplace—outside the country of use. In fact, in many cases a central purchasing organization, perhaps in England, may procure parts from a vendor in Japan to be shipped to the United States. Prices for goods purchased overseas usually will be quoted on the basis of a specific stable currency exchange rate at the time of delivery. For example, parts from an Israeli company probably would be priced in U.S. dollars. Transportation and customs tariffs are additional and should be included in analyzing the purchasing decision. Delays in customs are one of the more important considerations, and generally depend upon which country the goods are coming from and into which country they are going. Shipping parts into Spain, Italy, and Mexico is a present problem that can result in many weeks, and even months, of delay. Setting the reorder point higher to provide for safety stocks may be necessary in case of customs delays.

Packaging is also a concern in shipments. Several service companies have had success shipping large parts that obviously have little universal value on pallets covered with clear shrink packaging. Their visibility seems to make fork truck drivers and other handlers more careful, and reduces damage. On the other hand, goods that would appear to have some universal value should be shipped in containers that have only the part number displayed. Obviously a package emblazoned with the words "microprocessors," "television components," or even "flashlight batteries," would be a target for thieves. This problem, of course, is not unique to the international marketplace. If you buy a high-fidelity stereo from Sears, it probably will be in a plain brown box marked only with the stock number.

If an organization is starting to operate in the international marketplace, it should solicit advice from friendly service managers who have already encountered the problems and solved them. Knowledge of what transportation method to use, which brokers and freight forwarders can be fastest at expediting, the paperwork that is required, and the words to be used can be very helpful. Free trade zones should also be considered. Make sure that all publications are so written that they can be translated accurately into other languages.

12

Order Entry
and Supply
(Out-Bound)

PARTS ORDERS

ERVICE PARTS EXIST PRI-marily to fill the needs of employees and outside customers. This chapter on receiving and filling orders is meant to cover situations that range from a mechanic who comes to the stock-room window, to the field engineer who telephones that parts are needed in a remote location, to the customer or third-party service organization that wants to buy a part. The general flow of order entry and supply is diagramed in Figure 12-1.

Recognition of the need for parts comes in many ways. Many customers and field service representatives will have their own reorder point established to trigger orders before panic sets in. Others, because of lack of knowledge, negligence, lack of money to invest in parts, or unexpected use, will place emergency orders that need special handling. If an organization supports external customers and personnel, a regular mail/telephone/Telex/facsimile order system should be developed. Mail can be appropriate for routine orders that are not time sensitive. It is generally less expensive than the other media. Telex and facsimile have the advantage of written numbers and identifiers that reduce errors that may creep into a telephone conversation. A telephone customer service order entry function is always necessary as a backup to the others where time, availability of another medium, or the need for two-way conversation is important. Many organizations provide a toll-free 800 area code telephone number that allows outside personnel to call in their orders without incurring long-distance telephone charges. It should be mentioned, however, that the 800 number charges are very expensive and generally result in longer conversations than if the customer is paying for the call. Nuisance calls will also be placed over a free telephone number but may be eliminated if the caller has to pay directly. Wide-area telephone service (WATS) lines can be a marketing competitive advantage, but they should be carefully evaluated to assure that benefits exceed the costs.

A scheduled routine for order placing can be effective. For

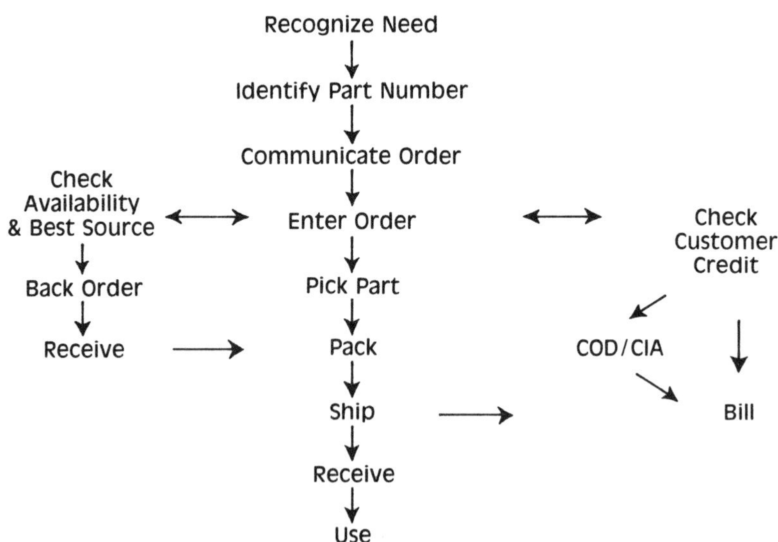

Figure 12-1   Order Entry and Supply Cycle

example, all orders for the week should be placed by Wednesday noon to be picked, packed, and shipped on Thursday morning. If field tech reps submit service call reports, they should indicate parts used. If those parts used are on their authorized stock list, opportunity should be provided through a check mark on the written form or a verbal response indicating that the tech rep wants that part automatically resupplied. A positive indication for a resupply is preferred to automatic resupply unless indicated "not wanted."

ORDER PRIORITIES

Most service parts systems will have at least two priorities for orders—"Routine" and "Emergency." Some add a "Stock" classification for filling reserve supplies. The emergency classification (sometimes called Urgent, Ship Same Day, Expedite, or Rush) should be used only in cases of extreme need, where equipment would be shut down as a result of need for that part. Emergency classification should be used on fewer than 10 percent of the orders. If the percentage is greater, then the system is being abused and the special-effort cases will overwhelm normal operations. An increasing use of emergency orders indicates reduced confidence in normal resupply, and the need for management attention.

Several rules regarding emergency orders will be helpful. First, any emergency order received at the ordering location should be installed within four hours. If it takes longer to get it in, then the need was obviously not of emergency classification. Second, the cost for special emergency handling should be recognized. Will-fit parts should be procured locally if the quality, cost, and speed of acquisition can be better than through the emergency channels. For example, fuses that can be obtained at a local electrical store should not be emergency ordered. Fastener hardware should be routinely replenished by the service representative well ahead of need. Numerous emergency orders for this type of item indicate a need to monitor routine replenishment better. Third, not more than twice the true emergency quantity should be ordered at the same time as the emergency part. If, for example, the reorder point on lasers is one and sudden demand causes stock-out, then no more than two generally should be ordered on emergency. If a special need for more as a result of reliability failures can be detailed, then more can be ordered; but allowing this as a normal situation will abuse the classification and special handling of emergencies. On the other hand, it does make sense to pick, pack, and ship two parts as emergency instead of one as emergency and the other through normal channels. Large quantities of parts ordered emergency should raise management suspicion that the system is not being properly supervised.

AVAILABILITY

While the ordering customer is still on the telephone, two checks need to be made. The most important concerns the availability of the requested part. The other is of the customer's credit standing. Availability should automatically be indicated by any computer program used for order entry or internal parts transfer. If an interactive on-line computer system is not available, then a printout that at least shows availability status as of the night before, or a stock record card that is updated with every order, should be at hand. If we are talking about a one-person order entry and issue function, then a card can be adequate. Even with three or four people, a rotating tub of cards can be satisfactory.

For internal availability additional information should be on hand for "On Reserve," "On Order," and "Back Order." The additional information gained from knowing what parts are on reserve will allow them to be diverted to an emergency situation, with management approval. Parts held on reserve are normally not under separate lock and key, but have been designated for a work order or service call. Often the stockkeeper will have picked those parts and have them assembled in a special location ready for pickup. If those are the only parts available to meet a high-priority need, then they can properly be used for the more important job and replaced when available for the lower priority one. The "On Order" status lets people know that the parts order cycle has been started. Sophisticated programs can even show how many parts are on which purchase order and the expected date of delivery. Note that a re-order point calculation should include both parts on hand and on order to avoid triggering additional reorder points signals every day until the new parts arrive. At the same time that calculation should indicate whether demands exceed both what is on hand and the number expected in, and what the timing will be.

Parts lent to another organization are easily accounted for as "On Hand" with that other organization. "Back Orders" may be used as a category, although it is not vital. That status of "Back Order" does convey the knowledge that the part order proceeded to the supplier and was acknowledged, but that the supplier is out of stock. From the central service warehouse perspective, a back order usually denotes an unfilled customer order that will be filled as soon as materials arrive. A replenishment is for materials to add to central stock against which there is no customer claim. Items on loan should be placed in the open order replenishment system with the appropriate return or due date. Of course, accounting for the loan item is separate. Again, knowing the expected date of delivery is important. Figure 12-2 shows a typical VDT screen layout for parts availability. Figure 12-3 is an external order entry screen that shows addresses, credit standing, and other information necessary for serving outside customers.

CREDIT

If outside customers are involved, then concern must be given to the ability and willingness of such customers to pay for the parts being ordered. While a service parts organization normally reacts to credit situations that are determined by a financial accounts receivable organization, service management should understand what is going on so it can play its role in intelligence gathering and enforcement. Naturally, if service sends out invoices and is responsible for payment collection, then their paychecks may depend on getting the bills paid promptly. If

```
           4 C     P A R T     A V A I L A B I L I T Y

PART #:   21544            DESC: CONNECTOR ,ELEC, 6 WIRE, MALE, FOR THERMOSPLIT
ENTER ORGN(O)/STKR(S)/EMPL(E):O#:123          NAME: CENTRAL SERVICE

                                          ON     ON    ON    BACK
     ID           ORGN/EMPL       BIN     HAND   RSRV  ORDR  ORDR
     ALL          GENERAL MAINT            48     8           144

11111             ARK
22222             BROWN
33333             BROWN                     8
44444             JAMISON                  16     8
66666             VINCELLI                  8
77777             MENG                      8
STK1              STOCKROOM 1      1B       8                 144
```

Figure 12-2   Parts Availability

```
      4 D      SELL   AND   SHIP   PARTS

ORDER #     498   DATE: 3/30/84 TIME 1645 RQSTR:JIM BACKUS    PHONE(716)243-0529
                                                                EXT  2435
           SHIP TO:       CREDIT  BILL      BILL TO:                  RELIEVE:
C/E/S:  C #:222                 C/E/S:  C #:222        E/S:  S #:11111
NAME:  WHITCO CHEMICAL          NAME:  WHITCO CHEMICAL        NAME: CTL STOCK
       PLANT MAINTENANCE               PLANT MAINTENANCE
ADDR:  BLDG 1, LINE 1          ADDR:  BLDG 1, LINE 1
                                                      PO#: 235-5
CITY:  BRADFORD     ST:PA      CITY:  BRADFORD    ST:PA
ZIP :13624-                    ZIP :13624 -
    PN           DESCRIPTION          QTY   UNIT $     EXTD $        COMMENTS
2222        OIL,MOTOR,10W-40,HD        5     1.25       6.25
3333        FILTER,OIL,DBL,GMHD        1     4.30       4.30


              TOTAL PARTS     2  TOTAL UNITS     6   TOT PARTS $    10.55
PRIORITY 3  SHIP VIA:  UPS    MAIL     PICKUP X  WT  2-8   TAX       $     .74
                       OTHER                              SHP/HDL   $
SHIPPED:C/P C DATE:  3/30/84 TIME:1700   DOC#:  WB455     TOTAL     $    11.29
STRIKE NEW-LINE FOR NEW ORDER - OTHERWISE PREV-MENU OR NEW-MENU TO EXIT
```

Figure 12-3   Order-Entry Screen

parts support is furnished as a follow-on to a capital goods lease or sale, then credit information should already be available. The "how to" of establishing credit arrangements for a customer are beyond the scope of this book.

Initial parts orders may be received from someone who has never been heard of before. This is very common in the case of durable goods such as home appliances, small sterilizers, and machinery that may be sold and easily installed at a new owner's location. The first knowledge a service organization may have of this is a telephone call or letter saying that the potential customer has the equipment and needs some parts. Obviously, being able to supply those parts quickly with a minimum of fuss can establish a profitable future relationship. Several organizations have been encountered whose credit departments take three to ten days to approve credit for a new customer, who meanwhile is anxiously waiting for an inexpensive part that is said to be readily available. Considering the cost of credit checking, a limit guideline could be established at about $100. If the order is for any amount under $100 and the customer appears legitimate, the order could be shipped immediately without a credit check. It naturally would be followed closely to assure that the customer does pay, and then that payment is recorded as an indication for future credit.

Higher order costs and orders to questionable or poor paying customers should be sent cash on delivery (COD) or cash in advance (CIA). Letters of credit (LC) and similar "cash against" documents are useful for foreign orders. COD can be arranged through most courier services such as United Parcel Service, Federal Express, Purolator, and the U.S. Postal Service. Those collection costs add to the price the customer must pay. The amount to be collected will consist of the cost of merchandise plus shipping plus COD charges. Some agents require either cash or a certified check for the order. The usual precedure is to tell the customer the exact amount and have the customer mail the check. When the check arrives, the part will be sent. Customers may pick up such parts in person with check in hand. If a tech rep is required to install the part, the customer can be notified what the call will cost for both parts and labor and agree to have a check waiting when the tech rep arrives. This should be done only in exceptional circumstances, as it is definitely preferred that field personnel not handle money in any form. This is a precaution both against temptation of the personnel themselves and against the possibility that thieves might try to rob someone they thought was carrying money. An invoice/receipt already filled out should be sent with the parts and a signed copy returned. These financial precautions may seem stringent, but they are intended to prevent illicit financial omissions or commissions. Service parts management is a business.

A manual system with a card for each customer, or even for each piece of equipment at a customer's location, should have a place for the credit to be indicated. Small orange and red adhesive dots can be useful for indicating CIA and COD status. Those dots can be applied when the credit people say it is necessary, and can be removed when the problem is cleared up. The customer order personnel should be able to say, "Yes, we have that part in stock. However, your credit standing is cash on delivery. Will you be prepared to pay the delivery person when it arrives?" Credit problems are not reserved for small companies. Many large hospitals, government organizations, and even divisions of "Fortune 500" companies have been put on restricted credit until all obligations are promptly paid.

PICK, PACK, AND SHIP

These three functions will be combined under one person in small facilities; or, in large facilities, may require many people. They may be done manually or in large part automated. Self-contained parts rooms are available that can be operated by a single person who keys the part number into the computer that guides the robot to the correct location, retrieves the part, and brings it to the counter. Those automated installations are most justifiable in locations that support high turnover of small parts, such as in a machine tool crib that supports a large factory. The same principle is in use in large warehouses built around retrieval and movement devices such as narrow-aisle fork trucks that can reach hundreds of feet into the air and lift several pallets at one time.

When an order is received at a parts location, it is best displayed in printed form. Planned orders can have their parts transmitted in advance so the parts can be staged in a location ready for quick pickup. The pick list should be printed in location sequence so the picker's route through the racks is as short as possible. In a manual system with under, say, a dozen parts on a typical list, the picker can look at the locations and write on the list the sequence 1, 2, 3 . . . . In a computer system, the program logic can sort the pick list into the most efficient order. Having pick lists sorted this way is a mark of a good service inventory system, whereas systems developed by inexperienced persons often are listed in part-number sequence. Very sophisticated systems, particularly in mail-order retail centers such as L. L. Bean can even consider the volume and weight of each order and direct the picker as to the size of cart to take and how the containers should be arranged.

When the picked parts are delivered to the packing area, a quality control inspection should be performed to assure that all ordered parts are present and in good condition. This check need not be very time consuming, but it is vital to the field personnel on the receiving end that they have the right parts to satisfy customers. Accuracy of parts picking is usually a function of the personnel involved. Inaccuracies in parts picking normally point to specific pickers who are in need of improved motivation and training. A visual display, possibly as simple as a chalkboard, can be a very effective announcement of each person's progress in number of orders picked, number of line items, and accuracy. Posting that information publicly provides a goal for everyone to strive toward. Persons who are below the norm will generally try hard to equal their peers, and there will be some who continually try to improve on their previous performance and strive for 100 percent accuracy.

Packing and shipping normally will be combined so that parts are prepared for the most suitable means of transportation. Protection of the parts while keeping weight at a minimum is the main constraint. Packages under 50 pounds and 22 inches in length may be shipped by USPS, UPS, or one of the many transportation services. Larger sizes and weights generally must be shipped by common carrier truck. If there is a large volume of packaging, advice on the best means can be obtained from professional packaging consultants or manufacturers' representatives. Sheets of air-bubble packing, plastic pellets, and even expanding foam are available for packing parts. Special attention must be given to electronic components such as PCBs. Instruct all personnel who may handle these components about the damage that can be caused by static electricity and how to prevent it. Microelectronic devices should be kept in individual, nonconductive packaging whenever possible.

SHIPPING

The economics of shipping depends on how much the package weighs, how far it has to go, and how soon it has to be there. Traditionally heavy equipment to be shipped long distances at the lowest possible cost would move by a combination of truck and ship. Increasing pressures from financial inflation and need for increased productivity now often put that same shipment on an aircraft for delivery in days instead of weeks or months. Many shippers provide combinations of truck pickup at your door and airplanes to fly the parts to the airport closest to the customer, where they will be picked up by another courier truck and delivered directly to the customer.

A service parts manager must be continually on the alert as to changing services and prices offered by competitive shippers. Locations within 400 miles may be served best by truck, whereas a combination of truck and airplane is necessary for longer distances. For low price and delivery to rural locations, a package sent by bus and picked up at the other end may be the best method. For local deliveries, a taxi or even the service organization's own van may be effective. Today a package under two pounds will be shipped anywhere in the United States for under $10 by the U.S. Postal Service. If notified by 3:00 p.m. in a metropolitan location, it will pick up the package by 5:00 p.m. at no extra charge and guarantee delivery the next day or refund the charge. It will ship packages of up to 70 pounds. The charge increases with both weight and distance. In the past, service organizations with essential accounts have designated a person to get the part from the stockroom and fly with it to the destination customer. This is expensive both in terms of the transportation involved and the escort's wasted time. It does, of course, get the necessary part to the location in a hurry. That same function can be achieved by companies such as Air Courier, which will send a messenger to pick up the part, get it on a specific airplane, and have it met at the other end and delivered directly to the customer. The service is expensive but less so than your own person doing that job, and the company is familiar with the best routing and schedules involved.

Consolidating shipments to once a week, or even less frequently, is very cost effective. It means packages can be consolidated by weight and size and shipped in the most economical fashion. That is, of course, a trade-off against carrying larger inventory quantities for safety stock. One expensive part on the shelf that can be quickly resupplied when used may be much better than an average inventory of five of those same parts with normal resupply only once a month. A smart manager will keep a separate list of emergency shipments and review them frequently to assure that the expediting system is not being abused. Too few emergency orders may mean that field inventory is excessive and money is being tied up in truly "spare" parts that are not turning over very fast.

FINANCIAL RESPONSIBILITY

An issued part usually also will involve a financial transaction to invoice a customer, or at least relieve the part cost from the inventory records. In the simplest form, that of a central stockroom that issues parts for internal work, the accounting may be no more than a slip of paper dropped in an "Issued" box or a line entry in a log book. There is no need for extensive bookkeeping if proper reordering and controls can be handled simply. As more money is invested in parts and more people become involved, planning and controls must become more sophisticated.

When an individual comes to the stockroom to get parts, a transaction record like that shown in Figure 12-4 should be made. This could be a line in a log book or an entry on a computer screen. The important elements are to know:

```
                        P A R T S     I S S U E

STOCK KEEPER ID#:   1234        NAME:   BROWN                    CHARLES
STOCK                       MEAS                   FROM                        TO
NUMBER      DESCRIPTION     UN    QTY   CODE        LOCATION      CODE      LOCATION
123434    SWITCH   ,DPDT    EA     1    S1          A21B3         3232
641118    GASKET   ,HEAD    BX     2    S1          F13A          3544
212431    SEALANT  ,SILI    EA    100   S1          F23A2         3544
311118    GASKET   ,HEAD    BX    10    3544                      S1        F14A
```

Figure 12-4   Parts Issue from Stockroom

1. The person who issued the part.
2. The person who received it.
3. Work the part is intended for.

The same information should be used for returning a part to stock. In principle, a part could be transferred directly from one person to another; however, in practice it is best to have the stockroom involved in the process and receive the part from one person and issue it to another. If direct person-to-person exchanges are permitted, valuable parts may disappear and the alleged receiver can say, "I didn't get that part!"

If all inventories are carried on line, then the inventory quantity can be transferred from the central stockroom to an individual's own stock inventory, and then further transferred to the specific customer or equipment when the call report is completed. If the customer is to be charged for the part, an invoice should be generated. The best source for that invoice is the stockkeeping financial function, although it may be given to the customer directly by the field service person who delivers or installs the part. Under any condition the invoice should be prepared promptly on delivery or shipment. Cash flow is important. The psychology too is desirable of getting a bill very soon after the satisfactory service has been provided. And especially if the invoice is for a large amount of money, the sooner it is received after the restoration of uptime, the more willing the customer will be to pay the charge.

Some organizations have their field service representatives present the customer with the invoice when the service is completed. Copies of the invoice then go back to headquarters for accounting, but the customer is expected to pay the presented copy. The problems with this are that the field service rep must (1) carry up-to-date costs for every item, (2) be accurate and businesslike in presentation of full charges, and (3) have the time and skill that are necessary to prepare and present an accurate invoice. It is generally preferable for the service rep to fill out the service call report and allow a central billing function to prepare and mail the invoice and worry about collection. Naturally, a one-person service organization is going to do everything and can prepare the bill most efficiently on the spot and present it to the customer for prompt payment.

To avoid any confusion over payment timing, a specific statement of when the invoice should be paid is helpful. A useful format is terms: "Net 20. Please send payment to arrive at Service InfoSystems, Inc., by September 12, 1984. Interest rate of 1.5% per month is due on any amount unpaid after that date." That simple statement makes it easy and direct for a customer's accounts payable function to pay the invoice on time. Regardless of whether or not a part is being charged, a receiving person's signature should be obtained on the issue form. That proof of receipt should be entered on the transaction log. The signed receipt should be retained as long as proof may be necessary. For a central facility, that may be only a day or two. For a field service time and materials job, it may be 30 or 45 days until the bill is invoiced and paid. It should not be necessary to send a copy of the signed receipt with the billing invoice. Reduce paperwork as much as possible and use electronic communications in place of paper.

RETURNS

Returns of unused parts to stock by internal personnel should be encouraged. The accounting mechanism should relieve the person or work order of those parts and credit them back to the stock inventory. Quality assurance can be built into those returns by assuring that personnel know the same part may be coming back to them in the future.

Quality of returns from external customers is a concern. Parts should be accepted for return only if good. An unopened package should be acceptable, as should a mechanical component that has obviously not been used or damaged in any way. Electrical components, on the other hand, rarely should be accepted for returns. Most service organizations refuse to allow returns of any electrical or electronic parts, because such items can be damaged by an electrical deviation and the fault may not show up until later.

A restocking charge of 10 to 20 percent is typical for those parts that are accepted for return. It does cost the supplier for the effort necessary to inspect the parts, put them back on the shelf, and adjust inventory and billing records. Special orders rarely will be returnable since they may be difficult to resell. Business practices for returns will vary depending on the amount of business done with the requesting customer. Like most situations in the service business, special attention should be given to the significant few, most important customers.

REDISTRIBUTION

In large supply systems, the concept of an excess redistribution parts pool may be employed. Excess parts from a field service organization are returned to one or more excess control points. Those stock points are managed as redistribution pools. Orders from lower level field locations are passed electronically through the excess pool(s) before going to the central stock point. A good place for an excess redistribution point is a small distribution center. This concept has the advantage of allowing the excess to "burn off" in the field organization without impacting the central stock point. It avoids the quality problems that may exist when a field organiztion returns parts of questionable quality to a central stock point that makes retail sales.

Movement of parts among members of industrial or field organizations requires close control if knowledge of the location of parts is to be accurate. Top-level accounting may remain correct, but lower levels will deviate unless some tracking procedure is used. An "internal parts transfer" procedure is recommended. This is usually a formal system that requires all parts to be returned physically, or at least in data form, to the stockkeeping location, which then reissues them to the requesting person or organizations.

Procedures often allow the giving person to input data that transfer the part, without an acknowledgment from the receiver. This does open the possibility that people will cheat the system by saying they have transferred parts off their inventory, but those parts never show up on the alleged receiver's physical inventory. This problem will usually be corrected when the latter complains that the parts were never received; investigation shows specific persons to be habitual senders of "ghost" parts. Acknowledgment takes time and effort, but may be necessary to avoid illegitimate transfers.

Redistribution of parts at financial checkpoints should also be watched. Mid-December transfers from field locations to higher stock points help the field location to close out the fiscal year looking good financially, but can make the higher locations appear to be poor managers. This game playing is often overlooked because it does at least put parts at the higher level where they can benefit a larger group of potential users.

13

Pricing

$\mathcal{S}$ERVICE STRATEGY MUST consider that future products will be more reliable and therefore will require service less often and will need fewer replacement parts. On the other hand, labor costs are increasing at a faster rate than are parts costs, so both cost containment and speed of repair dictate a movement to parts instead of labor. These factors should aim management attention at the marketing of service parts, and special focus on parts valuation.

This information is oriented toward persons who must price for selling. However, it may help other service managers to understand how pricing structure operates so that better purchases can be negotiated. Cost generally refers to acquisition value. Price generally relates to selling value.

CONSIDERATIONS

Any organization that provides service parts must at least cover the cost of buying those parts and having them ready for their customers, whether these customers are internal or external. The base cost of goods is the build or buy cost of acquiring the parts added to the internal carrying costs plus order entry, picking, packing, transportation, billing, returns, handling, and information costs.

Most manufacturing functions that produce parts and transfer them internally to service organizations raise the transfer price of those parts above their direct cost to cover overhead and support functions. The amount of that internal transfer markup can be the subject of considerable debate. Many managers feel that interdivisional transfers should not include burden. It is, however, a good way to transfer revenues and profits from one organization to another. Some service organizations that are very profitable shift some of those profits to manufacturing by paying higher transfer prices for parts "bought" from manufacturing.

Try to look at the pricing question from the outside, as through the customers' eyes. Other considerations include alternate sources for the parts, competitors' prices, speed of delivery, expediting cooperation, payment terms, warranty, return policy, packaging compatability, clarity of information, and responsiveness to special requests. The ultimate pricing criterion is the value of those parts to the user who needs them. Service parts prices are considered elastic, which means that increasing prices will decrease sales, and vice versa.

MARKUP CATEGORIES

Every service busines operated for profit should add a profit contribution margin to the costs. The margins will be greater on some groups and classes of parts than on others. The main markup categories are shown in Table 13-1.

Table 13-1
Pricing Categories

Category	Markup Multiplier
Proprietary unique	4–7
Modified	2–5
Common/universal/will fit	1–2

The markup multipliers are applied to the standard cost of the part. For example, a unique PCB purchased from manufacturing for $250 (manufacturing cost plus transfer cost) may be priced in the range of $1000–$1750, which is four to seven times the cost.

Proprietary unique parts are those for which an organization owns all rights to manufacture. The designs may be unique or there may be a special manufacturing process involved. The outside appearance should indicate that these are parts that cannot be easily found in common sources. Electronic modules, specialized castings, and rare materials are often found in this class of parts. It should be noted that software also fits into this category, and has many pricing characteristics similar to those of hardware.

Modified parts may look like universal, off-the-shelf items. Because of this, efforts should be made to distinguish them by color, special mountings, and packaging. Differential packaging may include a warning that other substituted parts may be less reliable, or even unsafe. A valve with stainless-steel inserts instead of phosphor bronze, and an electronic circuit board with different timing are examples.

Common/universal/will-fit parts include common hardware, chart paper, fuses, inks, and switches. Many of these items can be purchased at a hardware store or electronics hobby shop. Service organizations should provide any and all parts that a customer may require. Certainly any of those items that fail while under warranty should be replaced without charge. Equipment service under contract should normally provide the parts also. Most customers prefer to buy everything from a single source.

Top management must set policy as to whether an organization will attempt to get all possible parts business, or alternatively will provide all necessary assistance to customers so that they can get their own parts. If it is possible to provide complete contract service and/or rapid availability of parts, then it is recommended that service try to sell all possible parts. If, however, the products are remote geographically, require few sophisticated parts, and the customer maintenance personnel are competent to acquire and install the parts, then drawings and specifications for the parts may be provided directly to the customer. This trade-off has long-term implications that must be thought out carefully before a final decision is made. If service realizes that customers are going to buy their own common items anyway, service should assure that the proper specifications for these light bulbs, V-belts, and bearings are in the product's replaceable-parts list.

PRICING SENSITIVITIES

Customers may be knowledgeable about the price that they would pay for parts from other sources. The news media have featured stories of parts for military applications that cost the government (and the taxpayers) many times what an identical part costs at a hardware store. Many common parts are advertised in mail-order catalogs, and supplier salespersons drop by to disclose how much lower their prices are than anyone else's. Therefore, the margin should be kept low so that the perception to the customer is no more than a small premium for your one-source superior service. Universal parts are equivalent to the bread, milk, and eggs in the grocery business that are purchased frequently and, therefore, are obvious pricing sensitizers to prospective customers. Items purchased infrequently, such as table salt, which also sells at a very low price, are not significant pricing items. Consumables such as paper are particularly sensitive and may be priced low, often barely above break even, to create the impression that other prices will be equally reasonable. It appears that a price differential of 10–20 percent is the threshold that sends customers searching for other sources of parts. Obviously the absolute price has influence, so that a 20 percent premium may be acceptable on a $5 part since a $1 difference is not worth a separate order. However, on a $500 part that 20 percent would be $100, and a less expensive source would certainly be considered. Special consideration should be given to pricing parts for obsolete equipment. Prices should be increased over time to cover carrying costs on slow turning parts.

A minimum price, or at least a minimum order value, should be put on parts. A minimum price of $1 is frequently used, so that even a 12 cent cotter pin will be priced at $1. That is reasonable if a typical order is just for one or two of those items.

If, however, many inexpensive parts are needed, such as a fountain dispenser service person who requires mainly O-rings and gaskets that should sell for 35 to 60 cents each, then a minimum order charge is more practical. Terms such as a $2 handling charge on orders under $20 should cover the fixed costs of packaging, with shipping additional. To stimulate consolidated larger orders, incentives such as free shipping for orders over $100 and additional volume discounts are effective. Emergency, "equipment down" orders disrupt a parts system. Most organizations charge a handling premium of $5–$50 or a percentage such as 5–10 percent additional for rush orders.

SELLING THROUGH DISTRIBUTORS

A multitiered distribution system may go from producer to wholesale distributor to retail dealer to end user. If suggested list prices are published, dealers usually price at these, although some sell for less, and in some industries dealers sell at a 5–10 percent premium above the manufacturer's list price as a premium for fast supply. The standard procedure is for manufacturers to sell to distributors at a discount off list price, and for distributors to discount to dealers so that each echelon can cover its costs and make a profit. Twenty-five percent is a typical discount, and the actual amount is probably based on volume, with high-volume purchases or purchasers getting larger discounts.

PROFIT MARGIN AND RETURN ON ASSETS

Pricing structure is usually intended to generate a specific gross profit on parts sales. Specific markup multipliers may be chosen for particular categories of parts to achieve a specified gross margin on parts sales for the period. The calculation is

$$\text{Percent gross profit} = \frac{\text{Sales revenues} - \text{Cost of sales}}{\text{Sales revenue}} \times 100$$

$$\text{Cost of sales} = \text{Transfer costs}$$

$$\times \left(1 + \frac{\text{Cost of operating the supply system}}{\text{Cost of material handled during the same time period}} \right)$$

The overhead cost of operating a supply system is often about 20 percent. In the calculation this is best applied as a multiplier of 1.20. For example, if parts revenues for a year are $500,000 and the direct cost of those parts was $375,000 plus 20 percent operating overhead, then the calculation is

$$\text{Percent profit margin} =$$

$$\frac{\$500,000 - (\$375,000 \times 1.2)}{\$500,000} \times 100 = 10\%$$

Return on assets is based not just on the profit margin, but also on the number of times a profitable sale is made. The formula is

$$\text{Return on assets (ROA)} =$$

$$\text{Profit margin} \times \text{Inventory turnover rate}$$

Thus, if the turn rate is 3.5 per year and the margin is 10 percent, then the ROA is a very nice 35 percent. Note that the actual average inventory value was probably under $110,000. If a $200 part was sold for $20 profit margin, then another was ordered to take its place. This happened an average of 3.5 times during the year. High-turn items such as consumables can be sold at low margins. For example, a package of chart paper might cost (total of direct plus overhead) $14.55 and sell at a 3 percent markup for $14.99 ($0.44 margin). If turnover can be 12 per year, then the ROA is 36 percent or $5.28 on the $14.55 invested in assets to sell. For most industries it is far less, as shown in the Table 13-2.

PRICE CONTROL

Parts records should include costs, and may also include list price and our price, if different. It is relatively simple and

**Table 13-2
Industry Margins,
Turnovers, and Returns on Assets**

Industry	Profit Margin	Inventory Turnover Rate	Return on Assets
Agricultural equipment	4.4	2.4	10.5
Automotive equipment	4.0	2.3	9.1
Chemicals and allied products	3.3	3.1	10.2
Construction and mining equipment	4.9	1.9	9.3
Electrical supplies	3.7	2.8	10.5
Electronic parts and equipment	6.0	2.0	12.2
Industrial machinery	4.5	2.2	9.9
Professional equipment	4.8	2.4	11.5

desirable to construct a computer program to recommend prices based on desired margins with minimum and maximum guidelines. Retail prices should be monitored for their relationship to the transfer cost. Transfer costs tend to vary over time. Left unattended, fixed prices could fall behind cost increases and actually become less than cost.

Do not send cost information to field people. There is too much risk of it getting into unfriendly hands. Every organization that sells parts is concerned with quoting and invoicing the correct selling price. Two methods are typical. One provides a parts list with prices to all field personnel, who are then expected to put the proper price on the invoice. The other procedure is to have field personnel list only the part number and enough description to verify that it is correct on the service call report. The headquarters function, probably computer automated, applies the correct price, prints the invoice, and sends it to the customer. Either method can work, but each has advantages and disadvantages in specific situations. Field personnel will rarely keep printed price lists up to date. Somehow the latest revision is rarely in the book when it should be. Microfiche is a good medium for communicating prices to the field.

Knowing the prices of parts should help field personnel manage the territories better. For example, the facts should be available to decide whether it is better to replace a part at a cost of $35 or spend an hour trying to repair the problem. On the other hand, looking up prices on a list consumes time that might better be spent doing technical service. If prices are entered by a field person, then headquarters staff can audit them and the accuracy of two people is probably better than one. In the future electronic communications will probably eliminate the need for printed or filmed price lists. Instead, accessing the headquarters computer to report the service call data will allow the computer to respond with accurate prices. Likewise, a portable computer can provide prices, availability, and even invoicing at the customer's location.

14

Models

CRITICAL FEW

ERVICE PARTS MANAGE-
ment uses generalized models very effectively as tools and
methods to facilitate the management process. Some models, including ROP, EOQ, and Make versus Buy, have already been covered. It is not possible to concentrate on every problem or opportunity that presents itself. Attention should be directed to
the "critical few" and allow the "insignificant many" to fit naturally into the system without special attention. Pareto's principle in Figure 14-1 illustrates that point.

This is also known as the 80/20 rule because it is often found
that about 80 percent of all parts usage comes from 20 percent
of the line items. Probably another 20 percent of the line items
give 80 percent of the problems. Whether that number is pre-
cisely 80 or a few percentage points in either direction is not

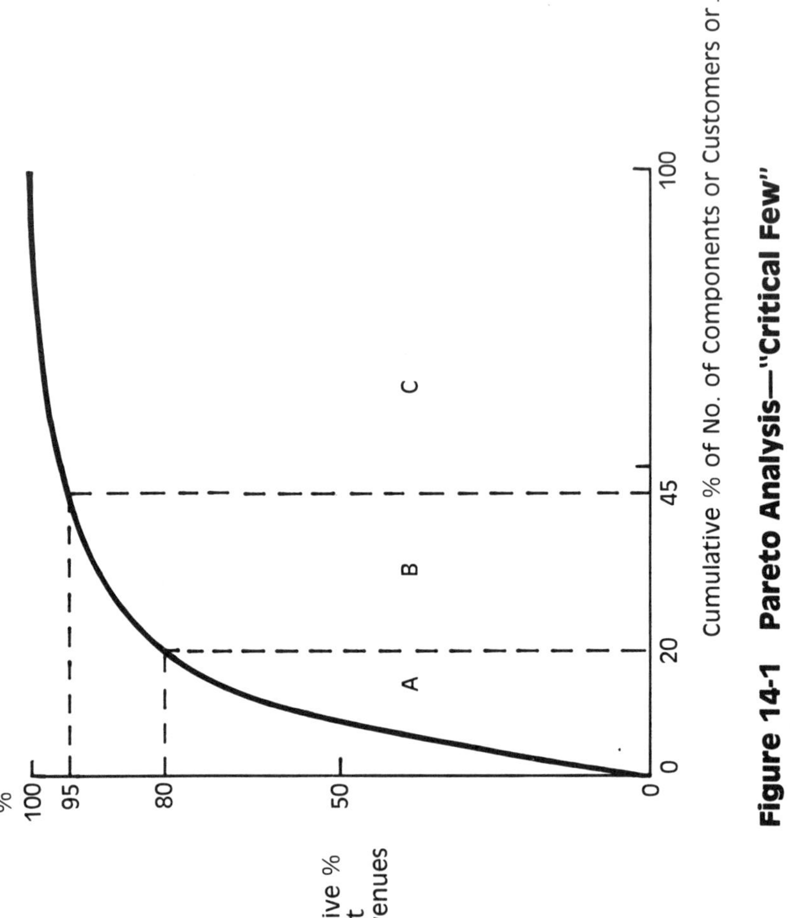

Figure 14-1 Pareto Analysis—"Critical Few"

important. The principle is observed often in many aspects of service management. A few customers generally contribute most of the business, a few pieces of equipment cause most of the problems, and so forth. An extension of the principle goes to 95/45; again showing that the vast majority, about 95 percent, of all activities is contributed by a minority, about 45 percent, of the items. The remaining 5 percent or so of the problems and opportunities come from the majority of items or events.

The critical few was clearly evidenced by a service organization that carried 6700 line items in its parts inventory. It initially had little idea what parts were used and what the turnover rates were. On analysis it was revealed that an 80 percent service level could be achieved with just 69 of those parts. A 90 percent level was achievable with 671. The remaining 6000 parts were there ostensibly to cover the remaining 10 percent. On further investigation it was shown that in fact they covered only about 5 percent and there were significant shortages remaining from parts that were not even stocked. The author's experience shows that this situation is all too typical. The challenge is to determine the parts that are being used and pay more attention to them.

ABC ANALYSIS

Manufacturing organizations for years have advocated ABC analysis based on multiplying quantities of parts used times the unit cost. Dollar thresholds are set so that parts that make up the top 75 percent of extended cost dollars are designated A, the next 15 percent as B class, and the remainder as C class. Figure 14-2 illustrates this point. Additional classes may be designated to include, for example, the following:

Class D—Parts with no issues in the past 12 months, but some within the past 24 months.
Class E—Parts with no issues in 25 months or more.

Other organizations use the following, which may also be separately coded as a commodity:

Class D—Common hardware and similar items that do not require individual item control. Reorder points are normally established by putting in a separate container a reserve/safety stock equivalent to at least the quantity that will be required during the lead supply time. When the prime stock is exhausted, the reserve container is opened and the reorder tag removed from the container and used to initiate resupply. These are often called summary stock items.

Class E—Generally used for bulk materials such as lubricants, solvents, and cleaning materials. These are usually purchased in large drums and individual small portable containers are refilled as necessary from the bulk supply.

Class F—Blanket order items such as stationery supplies that will be purchased on annual contract and accounted for by each order against the total commitment. Resupply time is generally very short so only small quantities of these items should be kept on hand.

Class G—Service contracts and similar support that are software items rather than hard goods. Examples would include instrument calibration services, equipment rental, and outside contract equipment maintenance. These are generally contracted for at a fixed rate, or on time and materials.

Class I—Insurance items. These parts are highly essential so management feels that they should be retained even though rarely used, since a lack could put equipment out of commission for long periods of time.

The manufacturing ABC inventory system that multiplies quantities used by unit cost is not extensive enough for service parts use. Attention additionally should be given to:

1. Demands
2. Essentiality
3. Controls

These factors may be calculated or coded and stored separately from the ABC classification, or they may be combined into a single rating.

Demand differs from usage as discussed earlier. It will be at least as great as the usage. If the additional unfilled demands are not recorded, then the consideration should be based on usage as an indication of the number of customers that will be satisfied. Given a $1000 choice between a single $1000 part and 1000 $1 parts, service people will usually choose in favor of satisfying 1000 needs with the inexpensive parts.

Essentiality, which was detailed in Chapter 3, is of value particularly for those highly essential parts that may be considered insurance items even though the usage does not justify having them in inventory. Control concerns the security that should be applied to the parts or equipment in question. Automobile parts, copper tubing, electric switches, and lasers are of concern since they are prone to theft and have considerable value on the black market or in outside-the-job activities. Hazardous substances such as mercury, poisons, and radioactive materials should also be controlled. Each part may have several codes assigned to it. Some are quantitative, such as ABC and demand class, and others are subjective, such as essentiality, commodity code, and control. Most quantitative codes can be calculated and assigned by computer programs, while humans must decide the subjective codes.

Small organizations may find it better to consolidate the impact factors into a single value class rating as outlined in Table 14-1. Obviously opportunities for varied analysis and flexibility are then limited, however, the KISS principle ("Keep it short and simple," or "Keep it simple, stupid") finds many virtues in consolidation. It should also be noted that consolidation to blanket orders and contracts has advantages in reducing the number of individual purchase orders required. Accounting of these specific events or shipments is still necessary, but paperwork is greatly reduced. A blanket discount is normally negotiated, with related savings in not having to contact the vendor to negotiate prices each time a purchase is contemplated.

CUMULATIVE—BY COSTS

The point covered in statistics is reemphasized that cumulative distributions are very helpful for determining the quantities and costs for achieving specific levels of service. As shown

Table 14-1
Consolidated Value Rating

Value Class	Demand Low	Demand Medium	Demand High	Unit Cost Low	Unit Cost Medium	Unit Cost High	Essentiality Low	Essentiality Medium	Essentiality High	Control Low	Control Medium	Control High
A Critical few			X			X			X			X
B Medium attention		X			X			X			X	
C Low attention	X			X			X				X	
D Common hardware			X	X						X		
E Bulk				X						X		
F Blanket		X	X	X		X				X		
G Services				Services cut across all factors								
I Insurance	X								X			

X = Main reason

in Figure 14-2, a cumulative distribution can be very helpful to point out the significant few, as in Pareto Analysis. For managing parts the first few screens or pages of printout as illustrated in Figure 14-3 will show the high-priced, high-demand items that should receive stocking attention. The remaining items will be of far less importance and should be further evaluated to determine whether they are being used and whether they should continue to be stocked.

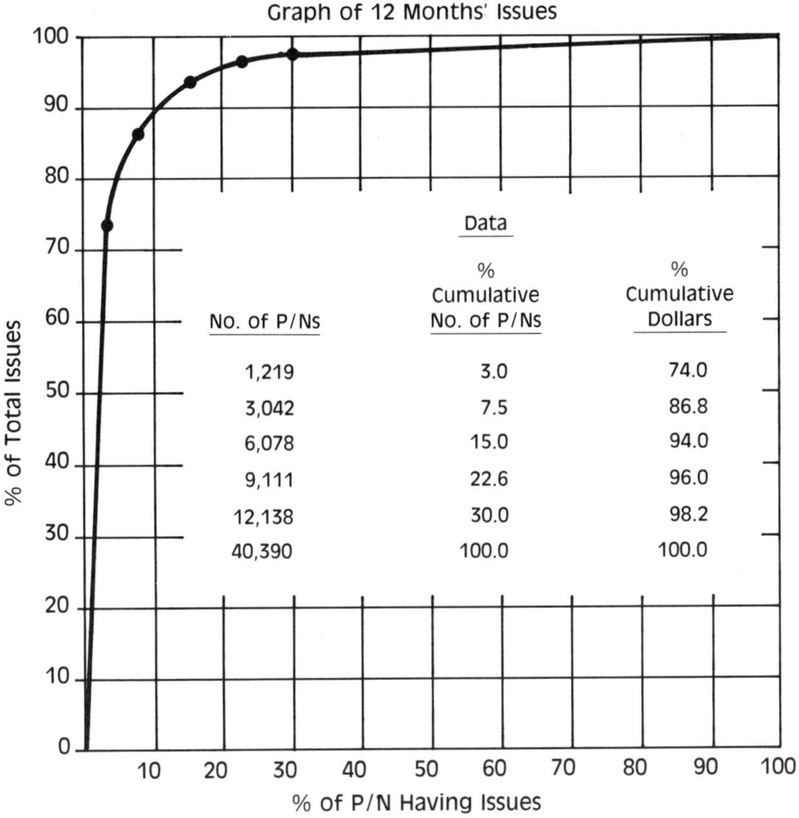

Figure 14-2   Example of ABC Analysis of a Central Inventory in a Service Parts Supply System

```
4 Q     ALL  INVENTORY  ON-HAND,  BY  COST

EMPL(E)/STKRM(S): S #:S123        NAME:MOTOR POOL STKRM    DATE: 5/ 5/84
ENTER SCREEN VIEW(S)/PRINT COPY(P): S
```

PN	DESCRIPTION	QTY	UNIT $	EXTD $	CUM $	CUM %	DOH	VL CL	ESN
8888	PRNT CK BD,MOTHER BO	5	345.00	1,725	1,725	86	50	B	2
9999	TRANSFORMR,110V,30W,	3	86.75	260	1,985	99	30	B	2
5555	FILTER ,AIR,TRUCK	4	2.30	9	1,994	99	40	C	3
3333	OIL ,PENETRATI	6	1.20	7	2,001	100	60	C	4
4444	FILTER ,MOTOR VEH	3	1.70	5	2,006	100	30	C	3
1111	OIL ,AUTOMOTIV	4	.65	3	2,009	100	0	B	3
2222	OIL ,CUTTING,S	1	1.25	1	2,010	100	10	C	3
405008	SPRING ,RELIEF,IT	2			2,010		20	B	2
405009	SPRING ,BOTTOM &		2.75		2,010			C	3

Figure 14-3   Parts on Hand By Cost

TURNOVER

Turnover is the measure of how often a given line item is used during a year. The calculation is based on issued quantity/average quantity on hand. For example, if 12 parts were used during the year and the average quantity on the shelf was three, then the turnover is $12/3 = 4$ times. Turnover could also be evaluated using cost as the base, but variable costs attributable to inflation and repairs make the calculations difficult and often erroneous. Quantity-based calculations are preferable. Naturally a high turnover is better than low. For example, if the turnover of a PCB is 1.5 times a year and the average quantity on hand is six, with a $9000 extended cost, then increasing the turnover to three times a year could support the same activity with three PCBs and cut the inventory in half to $4500.

Turnover calculations require large amounts of on-line memory to calculate in a computer system since the average quantity on hand must be retained (stored) for periods typically by month. At the end of each month, the quantity on hand would be averaged and that number retained for use in later calculations, usually based on a year. This is too much for most microcomputers, so "date last used" is often checked to detect low-turnover parts. Low-turnover items are a rich source for disposition analysis and return to the source or use in current manufacturing.

Turnover can also be viewed as weeks of supply:

$$\text{Weeks of supply} = \frac{\text{On-hand quantity}}{\text{Demand quantity/year}} \times 52 \text{ weeks/year}$$

Weeks-of-supply analysis is very useful in analyzing groups of parts, especially within commodity groups and classes. Since turnover can vary depending on the age of supported equipment, confidence in resupply, lead times, and desired service level, turnover may not be the main goal. Customer service level and economic benefit/cost are generally more important. Note that good management will get high turns as well as a high level of customer support.

For most service organizations, and certainly for anyone retailing parts, turnover is a necessary objective.

BREAK-EVEN ANALYSIS

Break-even (profit) analysis is useful for calculating how many of an item or how much of an activity must be accomplished or how many items must be sold to be financially viable. As shown in Figure 14-4, the three components of break-even analysis are fixed costs, variable costs, and revenues. The fixed costs are those expenses for facilities, personnel, utilities, and equipment that will be incurred regardless of whether they are used or not. Variable costs are those expenses that vary directly with the amount of activity for the number of items. Revenues are the money that comes in from repairing each product, selling each item, or accomplishing the activity.

Consider, for example, repairable service parts. A control module costs $300. We can repair each of them for $150 for di-

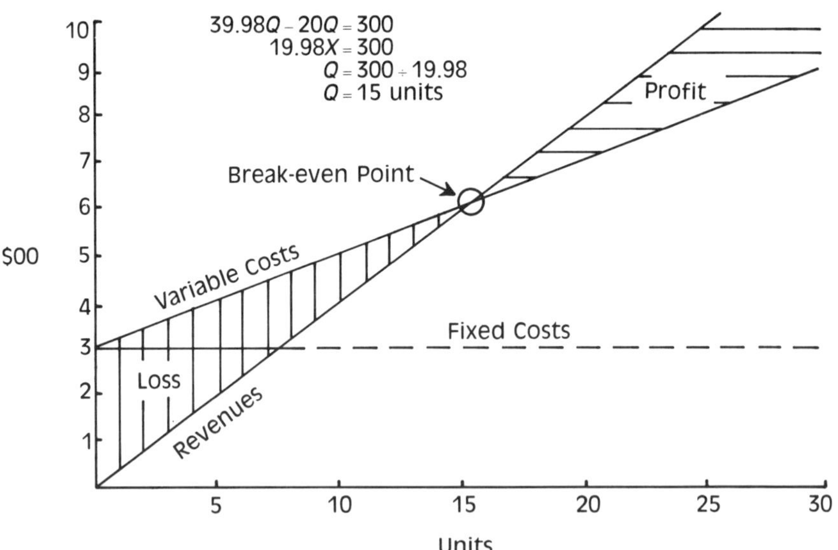

Figure 14-4   Profit (Break-even) Analysis

rect parts and labor, shipping, and related expenses. Failures are projected at 15 per year. A computerized tester is required, at a cost prorated to $4000 a year. As the facilities and other fixed costs to establish and operate the repair center can be shared with other support, this adds only $2000 a year. Should we repair the modules? The answer is yes, if marginal revenues exceed costs. Otherwise the answer is no.

The calculation is

Revenues
(Unit revenue × Quantity)
or
($300 × 15) = $4500

Costs
(Unit direct cost × Quantity) + Fixed costs
or
($150 × 15) + $6000 =
$2050 + $6000 =
$8050

Since the incremental revenues are much less than the costs, the venture does not make economic sense and probably should not be carried out. Additional questions should be asked, including: Is the capacity for repair vital? Can enough new replacement modules be obtained if the old ones are discarded? Are there personnel who must be kept employed? If the economics are close to even, those support questions may force the decision.

MATERIAL REQUIREMENTS PLANNING

Material requirements planning (MRP) is very useful for planning materials needs for production. The process starts with defining a bill of materials (BOM) for each item to be manufactured. There may be several levels of hierarchy in the BOM. For example, a heavy truck would be made up of assemblies such as body, engine, and transmission. The engine, in turn, would be made up of subassemblies, including the block, fuel injection, and cooling. The block would be made up of units including valves, springs, and gaskets. The BOM breaks equipment down to the lowest possible configuration that is encountered by the organization. For service parts management purposes, it is only necessary to define the bill of materials to the level at which parts are procured and used. For example, if a control module is purchased intact from a vendor for manufacturing, service needs to have a part number for the complete control module. The manufacturing organization needs to know the detailed contents, but service does not. In fact, service needs to have part numbers only for replaceable parts on equipment. Components that cannot be replaced even if they do fail should not clutter the service parts database.

It is very helpful to know the next higher assembly for every unique/proprietary part. If, for example, a specific integrated circuit is not available, then possibly the PCB is. If the board is not available, perhaps the complete module is. It is also helpful to be aware of any interchangeable parts, both from old to new and from new to old. Perhaps a 60-watt bulb is not immediately available, but a 40-watt or 75-watt bulb can be substituted. These bits of information all can help the system to be flexible and meet the end goal of customer satisfaction at minimum cost.

NOMOGRAPHS

Nomographs are a series of curves that may be intersected by straight lines at given values, which will indicate resulting values at the intersections of other related curves. Figure 14-5 shows a nomograph for calculating the number of parts necessary to provide a given level of service.

In the example we first need to know the total number of operating parts to be supported. Assume there is one per equipment and 20 equipments, so a mark is placed at 20 parts in column (1). Then the expected failure rate of the parts is marked in column (2). The responsible design engineer or reliability analyst should be able to provide the rate. Here we use 0.1/1000 hours (100/million hours). Use a straight edge to draw a line that connects the two marks. Then mark in column (4) the length of time the equipment is to be supported. Here the period is three months. Connect this point with the index in column (3) where our previous lines crossed, and draw a line to (5), which is $K\lambda T$. You then must decide the level of support desired and mark the probability (here 95 percent = 0.95) in (6). Draw a line from (5) to (6). Since our line crosses just below 8, the answer is to stock eight spares.

SIMULATIONS

Simulation is a form of modeling that processes the elements in a pattern similar to what would be expected in the real world, and evaluates the results to assure that proper systems, quantities, and resources are in place. For example, instead of assuming that parts are used at a rate of 24 a year or an average of two per month, the simulation would use the log-normal, Poisson, or other distribution that the demand is expected to follow. If causes can be identified, then they should be included. If the pattern is random, then a random number generator should be used to enter the values. Several passes through a simulation can usually point out where problems may arise. Obviously modern computer power is very effective for this purpose.

Simulations may also be done in physical form using three-dimensional iconic models to assure that a stockroom is properly laid out or that vehicles are loaded to the best advantage. The word *heuristic* means to operate as the human mind does. A heuristic computer program follows the same general steps that the human parts clerk or picker/packer would follow to do the task. The computer, of course, can calculate the results of many alternate approaches in a few seconds and select the best.

The most important requirement of any model is that it reasonably represents the real world.

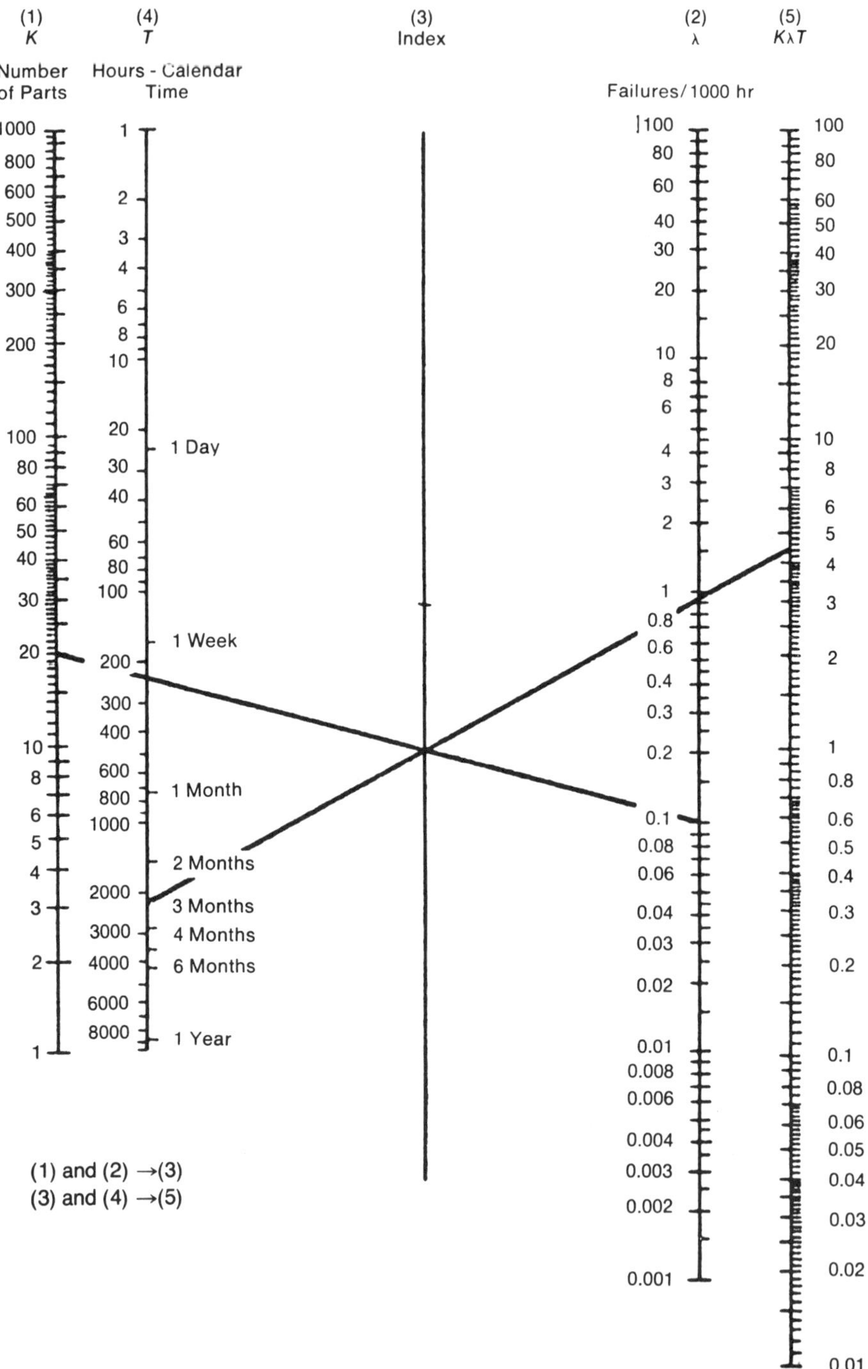

Figure 14-5A   Spare Part Requirement Nomograph (Sheet 1 of 2).

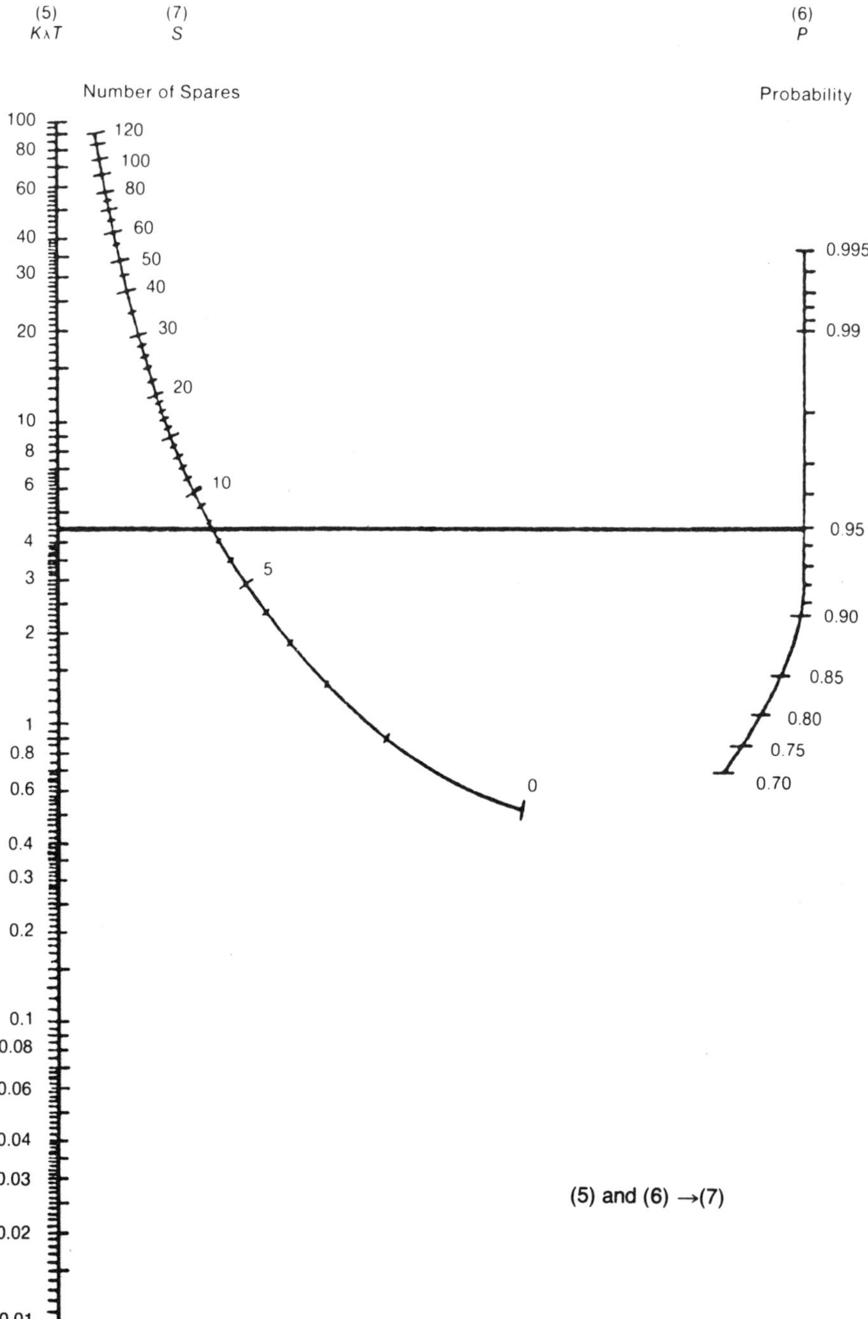

Figure 14-5B   Spare Part Requirement Nomograph (Sheet 2 of 2).

15

Repairables

REPAIR VERSUS DISCARD

MANY SERVICE PARTS HAVE the virtue that they can be repaired, refurbished, or rebuilt. That means their value can be restored by replacing damaged components, joining broken pieces, or reforming materials. This capability makes it possible to retain the materials and labor value that would be lost if the part had to be discarded. It also provides the capability for restoration for use in difficult situations where new parts may not be available but the means of repair can be.

A few years ago, maintainability and manufacturing strategy tended toward discard rather than repair. It is generally less expensive to manufacture parts that are to be discarded. The production process can use epoxy, welding, encapsulation, and other techniques that save money in production. Unfortunately they render the components unrepairable. Trade-off analysis often showed that was the correct way to go since there can be advantages to service by not having to train personnel in repair techniques or spend money on repair tools and facilities. However, there are three factors that work against success with a discard policy. First, there must be an adequate supply of replacement parts since every failure requires a new replacement. Second, international customs barriers between countries require extensive paperwork to ship parts back and forth. Most customs officials cannot tell a failed part from a new one, and thus the same duty will be charged on a defective part unless it can be proved that it is a specific replacement. The virtue of repairables is being able to keep the part within the same country and avoid having to pay customs on all the new parts being introduced for repair.

Third, adequate recognition was not given to the high level of motivation of technicians with regard to repairing parts. An electronic module would be classified as a throwaway, but instead of doing that, the technician often would take it home, troubleshoot the problem, purchase any replacement components at the local electronics shop, and restore the module to a usable condition. Of course, the fact that it was done on the person's own time was outside the economic analysis performed for repair versus throwaway, and this is positive factor, especially if the technician cannot get an adequate supply of new replacement parts. The only real problem is that the component used in replacement is often of inferior quality and the workmanship involved may cause unreliability and future failures. Pennies saved on parts' costs may result in dollars lost through additional service calls and customer dissatisfaction.

REPAIR TAGS

The most beneficial thing a service technician can do upon removing a part that should be repaired is to tag it for return, with the equipment it came from identified and the reason for removal detailed. If the symptoms can be noted, that will be a major help to repair personnel. Many parts returned for repair, often as high as 50 percent, are found not to be defective. As pointed out earlier, the parts may be good but were removed as one of many replaced during troubleshooting. The real problem may have been an edge connector, another of the removed

```
               REPAIRABLE PARTS RETURN TAG

Part Number:  86127        Noun Description: Powersupply

From Product: P 300         Serial Number: 31297B

Service Call #: 45338          User ID: 325

Problem with this part: Intermittent high voltage
after hot from long runs.


Employee #: 179        Name: Buzz Sawyer
```

Figure 15-1   Repairable Parts Return Tag

components, or software faults that appeared as hardware. There may be, in fact, a collection of problems or tolerance buildup in the specific equipment and circumstances that do cause a failure, but the same problems could not be easily detected in the laboratory or repair facility. The point is that any information that can be provided by the on-site technician will be very helpful in repair. A typical repair tag is illustrated in Figure 15-1.

PACKAGING

Any part that is designated as repairable should be packaged in returnable containers if there is any transportation involved. A automotive generator, expensive electronic modules, and warranty exchange items are examples. The packing material should be durable. It should be plainly marked to the effect that it is a reusable container for returning the repairable items, so that the opener does not destroy it. Inside the container should be the defective parts information tag; complete instructions for exchanging the new part in the box for the failed one; the return address label, prepaid postage, or other transport arrangements; and sealing tape. The part recipient should just have to remove the new part from the box, apply the return tag (with information entered) to the old part, put that part in the box, seal the container, apply the address and postage label, and put it in the mail. Arrangements can even be made for UPS to pick up the return package. An alternative that is preferred where practical is to return the item to a designated collection point for bulk shipment to the repair center.

If a plant or local repair situation is involved that precludes the need for transportation, then a return parts information tag may be the only item required. It is best to have that tag on new parts so that when the new motor or other replacement item is given to the user, the tag can be removed and applied, after the pertinent information has been added, to the return-for-repair item. The result of a well-thought-out and enforced return packing and tagging program will be a high rate of returns with helpful information, resulting in better turnover of repairable parts.

CREDIT

One of the questions usually asked is, "How much credit do I get for return of this repairable item?" Most organizations provide an exchange price that credits the reusable value of the returnable, defective parts. That figure is usually calculated over many repairs and considers scrap rate, shipping damage, and the range of repairs required, with the average value used for all returns of a part number. The value is usually 30–50 percent of the new part's price. Of course, the higher the credit, the more motivation there is to return parts.

Credit should be given promptly on receipt of the failed part. One way to credit for returns effectively is simply to invoice for the new part assuming trade-in. If the failed part is not received in 30 days, then a second invoice is sent for the credit value plus handling charges. Some organizations invoice the full amount and then give credit on a voucher when the returned part is received. That requires two invoices, whereas only one is usually necessary with the first option, and one invoice is better than two. Another alternative is not to send a new part until the defective one is returned, but that is slow and causes longer downtime.

LEVEL OF REPAIR

The factors that impel level of repair (LOR) decisions are quantity of products to be repaired, cost of each, skill required, equipment necessary for diagnostics and repair, and transportation time. The main levels of repair are as follows:

1. User
2. Technician on the spot
3. Local bench
4. Regional repair center
5. Factory or central repair

Often steps 3 and 4 are omitted and repairable parts that cannot be fixed on the spot are sent directly to a national center. It has already been emphasized that repair by the user is the best possible situation. Figure 15-2 shows the trade-off between costs and quantity, with the cost factor integrating delay times, required skills, test equipment, facilities, materials, and personnel.

The economic analysis is relatively straightforward. The equation minimizes the cost of repair and identifies at which location it will take place. This is simply a variation of the break-even/profit analysis illustrated in Chapter 13. A "cookbook" worksheet format is illustrated in Figure 15-3. A microcomputer spreadsheet program such as Visicalc or Lotus 1-2-3 is a good way to do LOR analysis.

ELECTRONIC GUIDELINES

Electronic modules such as PCBs are prime candidates for repair. When circuit boards are returned, they should be logged in to account for them as work in process. Then they should be tested, using information on the returned trouble tag as a guide to the defects and necessary actions. If engineering change orders (ECOs) are required, the changes should be installed be-

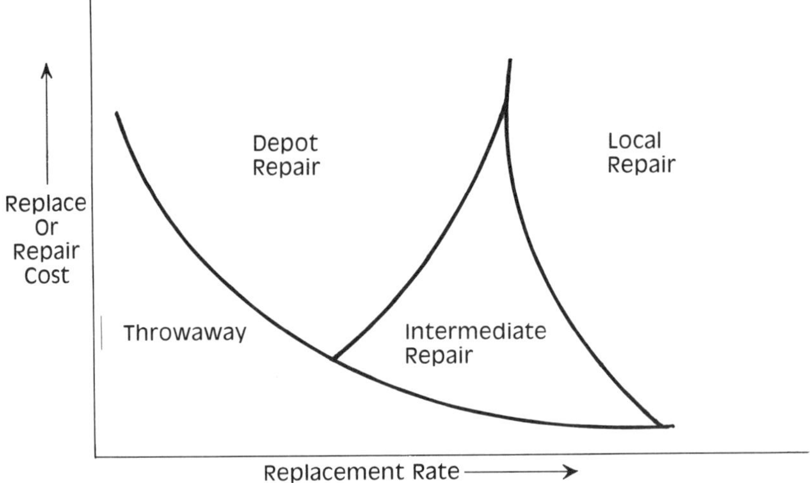

Figure 15-2   Level of Repair

fore testing. Why? The answer is it must be done anyway, the installation may eliminate the problem, and if the unit tests good, then repeat testing is eliminated. After the boards are repaired, time and materials should be logged against them so costs and repair performance can be measured.

Bar-coded part numbers and serial numbers help in data collection and analysis. Tracking components replaced, labor, problem/cause/action, and turnarounds will provide information to improve many situations.

On completion of repair, the boards should be marked to show that they have been through a repair cycle. One good way is a dot of colored paint or ink or a stick-on colored tab. If a defect was found and corrected, a yellow mark is suggested. If a defect could not be found, then use a red mark. If a board with a red mark comes in for repair a second time and a defect cannot be found, then the board should be scrapped. The reason for this drastic discard action is that there is probably a defect such as an intermittent electrical fault that may not be found in testing but does appear in the user's equipment. Further time should not be wasted trying to find the problem. The level of

Fixed costs:	**User**	**Technician**	**Local Bench**	**Region**	**Factory/Central**
Documentation					
Facilities					
Utilities					
Test equipment					
Tools					
Training					
Base salaries					
Variable costs:					
Quantity of repairs per year					
Labor hours per repair					
Labor rate					
Skill level					
Materials costs					
Special costs:					
Pipeline quantities					
Training					
Documentation					
Inventory carrying					
Other considerations					

Figure 15-3 Level-of-Repair Analysis

quality and the cost of service are enhanced by eliminating the board from further use.

Organizations may limit the number of repairs to be made before a part is discarded. Any limit should be determined by the physics of failure. With electronic circuit boards, the limiting factor is generally delamination of the connector circuitry. That results from frequent soldering and unsoldering. An inept soldering job can ruin a board in one stroke. On the other hand, carefully applied techniques may result in dozens of good repairs. Of course, if work is done on different areas of the same PCB, then the combined effect may be different than if all repairs are concentrated on the same area. Eight repairs per board is often used as an estimating guideline.

Another guideline that is valid for any kind of part or repairable equipment is that items valued at less than $100 should be discarded rather than returned for repair, unless they are not replaceable or no replacement is available and so the defective part must be quickly repaired. The exact cost can be calculated for each organization, but it generally costs at least $100 for handling, packaging, data, transportation, and testing of any returned item. It does not make economic sense to pay $100 to return a part of less value.

QUALITY CONTROL AND ASSURANCE

Repaired parts should be subjected to the same quality controls as any new part. As with new parts, quality must be designed and processed in rather than inspected in. In other words, the diagnostic and repair process must be done logically with quality and reliability in mind at every step. There is little value in finding defects after they have added expense to repair. Above all, it should be remembered that repaired parts are probably going directly into the replacement parts inventory. They rarely will be installed in new machines at a factory. Therefore, their first trial in actual functional use may well be as a repair part for a problem that is receiving the full attention of an angry user. It would be adding insult to injury if the replacement part failed too. Anything less than 100 percent acceptable quality of replacement parts will be much more expensive than the alternative of investing wisely to gain high quality.

MANUFACTURING VERSUS SERVICE REPAIRS

A question that is often debated concerns which is the best organization to do parts and equipment repairs. Considerations include the equipment, tools, and training required; the similarity between field parts and factory parts; volume of repairs; economies of scale; and possible/required turnaround time. If the field parts and assemblies are the same as those in the factory, then the factory should be considered. If field configurations are different, than a separate repair location should be established. Factory assembly lines are generally most efficient for high production. Service part repairs are generally a short job-shop-type process. A run of five or ten is typical. It is usually not effective to set up a factory manufacturing operation for a run of that size. Often demand for the parts must be given priority over repair process efficiencies.

Motivation is another major consideration. Management may say that manufacturing will be just as motivated as service would. But, as with most activities, people pay the greatest attention to things that benefit them the most. Manufacturing's business is to produce new products. If repairs are a task that does not add to their performance appraisals, then repairs will get less attention than they need. Also, service will be organizationally motivated to turn repairs around in a hurry and get them back to effective use. If it is best to have manufacturing perform repairs, then motivation through high transfer costs that make repairs "profitable" by covering overhead and contributing margin should help. Specific turnaround objectives should also be established so that, for example, no repair part is in house longer than two weeks, and the initial value plus repair cost does not exceed two thirds of a new part's cost. The manufacturing system frequently does not support the batch-oriented process of repair. Most batches are split into sub-batches, depending on the type of problem. The actual cost per batch of labor and material is needed to evaluate exchange credit and other economic values.

A materials requirements plan (MRP) should support the unique bill of materials for items to be repaired. The bill quantity can be stated as a percentage of the boards to be repaired. For example, an integrated circuit (IC) may be forecast at a 20 percent replacement rate. This can then be matched to the master schedule of expected returns so components can be procured for the projected repairs. This gets all components, both new and repaired, into the master schedule. The same quality standards should apply to reconditioned parts that apply to new ones.

PIPELINES

The word *pipeline* is used to describe the routine timing and quantities of parts necessary to support all elements of an organization. For example, a new product being distributed for the first time from the factory requires spare parts in the field either before or concurrent with installation of that equipment. If each distributor/distribution center needs 20 units, each local servicing organization needs five, and each servicing technician needs one, then the quantities can be determined simply by multiplication of the number of facilities each requires. For example, (3 distribution centers $\times$ 20 each = 60) + (60 local servicing organizations $\times$ 5 = 300) + (500 technicians $\times$ 1 = 500) = 860. That is the quantity that will be required in the pipeline to support initial service.

The timing of the pipeline quantities can be calculated on the basis of how long it takes for the parts to get from the factory through the distribution channels down to the technician who is going to install the equipment. The pipeline quantities will remain generally constant as a function of equipment volume and problems that necessitate part replacement. When parts are replaced on equipments operating in a plant, generally the technicians install their replacement parts and order others from the local center so that they are resupplied. When the local center reaches its reorder point, it orders replenishment from the national distribution center. And when distribution reaches its reorder point, it obtains more from the factory. Thus the lead time is not the time through the total system, but rather the time to resupply the part from the next higher stock location. Defective parts being returned for repair should go to the repair facility by the most direct means possible. Time is money, so investment in faster shipping or in speeding the repair process will often save in the end and increase profits.

The operation of the return pipeline is significant in many service parts supply systems. Field repair has the advantage of shortening the repair pipeline and the quantities of parts in it, but the disadvantages of high setup costs and potential quality problems. Central repairs at a depot or the manufacturer have the advantages of lower operating costs, better quality, and engineering change control, but require longer repair lead times. Field repair is generally more cost effective for high-volume but low-cost parts. Central repair is usually better for low-failure-rate but complex, high-cost assemblies.

The amount of time that parts spend in a failed status is of concern in any supply system. The failed-part turnaround process should be constantly monitored to assure that the minimum amount of material is in the failed status. The decision to repair a particular part should be made in view of the overall availability of inventory in the supply system. Excess parts need not be repaired prior to need. At the same time, manufacturing may be waiting for enough parts to accumulate before it has what it considers to be an economic batch, but service may have those parts on back order. An example of a failed-part turnaround process is shown in Figure 15-4.

REQUIREMENTS CALCULATION

Calculating the number of parts required is different for repairable parts than for throwaways. Once the initial pipeline quantity is calculated, demand is filled by turning over the repairs quickly and getting those parts back into good use. The time from user to repair location is a major element. With a captive service force, this is motivated by making it an item on the job description and performance appraisal. Economic trade-in value is the motivator for noncaptive returns. Ten calendar days is a reasonable return time, with 30 days maximum.

Once returned parts are in house, diagnostics, repair, quality assurance, and packaging can average less than a week if the need is great. Often modules gather in a corner until there are enough to set up the process. A good goal is to have parts go through the complete repair cycle in less than one month.

This means a 12-times turn rate; so that one part can, on the average, support 12 failures. If the repair cycle takes three months, then three parts would be needed to support the same failure rate. Obviously a faster repair cycle pays off in the need for fewer "spare" parts.

Table 15-1 shows a planning process that can be used for both new and repaired parts planning.

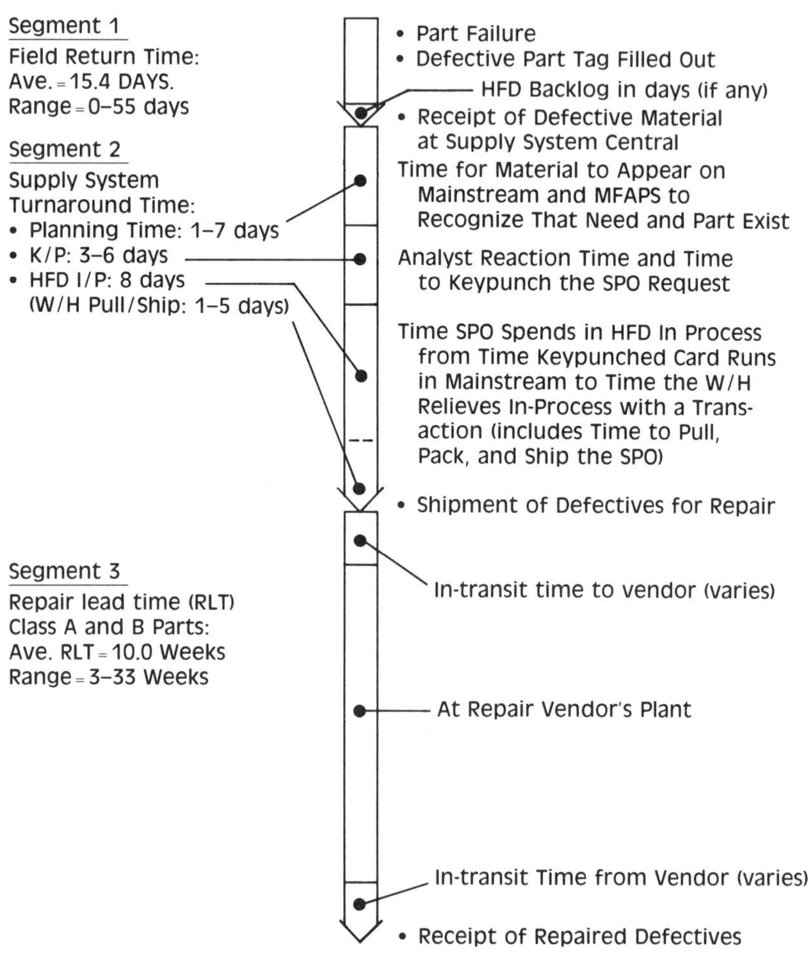

Figure 15-4   Defective Part Turnaround Process

Table 15-1
Time-Phased Planning

	Present	1	2	3	4	5	6
Total requirements forecast	25	30	30	35	35	40	40
Repaired (60% of prior month)	15	15	18	18	21	21	24
Need new	10	15	12	17	14	19	16
Receive Order (EOQ)			48			48	
Total on hand EOM	20	5	41	24	10	39	23

16

Computerization

Virtually everyone in the service parts busines is aware of the influence of computers on all facets of life today. Electronics wakes us up in the morning with beeps and music, turns on the coffee pot, opens the garage door, and even provides guidance to the ignition and other systems of the automobile that we drive to work. Once on the job, we utilize computer capability to analyze information rapidly. And, in the service business, many of the parts we are ordering and supplying are electronic. Forty years ago computers filled large rooms and were programmed by moving cables much as an early telephone switchboard was. Today even greater computing power is found in portable microcomputers that can be held in the palm of your hand.

JUSTIFICATION

Service parts management can benefit greatly from the capability of a computer to store, recall, and perform calculations on large groups of numbers, very accurately and reliably. Justifying computerization is a concern of many service managers. Inventory records can be kept in a person's head, in a notebook, on cards, on strip charts, or in a computer. A small organization with relatively few stock numbers, say under 500, can probably manage the operation just as well on cards as with a computer. Certainly the initial investment is less. However, as the numbers of parts stocked, the dollar value, and corresponding need for control increase, a computer will be of benefit. As a general rule, activities should be computerized if they are repetitious, require accurate calculations, and involve many manipulations of information. There is very little in service management that cannot be computerized effectively, given enough attention by intelligent service managers with enough money, time, and computer resources. It is those resources that present the major stumbling blocks, since there are rarely enough resources available to do what service managers can dream up. The challenge then is to assure that the benefit will at least equal, and, if possible, exceed, the cost.

Unfortunately, few organizations know how good or bad they are until a computer system is implemented. Experience shows at least a 10 percent reduction in the number and cost of parts necessary, at least a 10 percent improvement in part availability at the same time, and major labor savings and equipment productivity increases due to having the right part at the right place and time. As an example, one "Fortune 100" company with 18,000 parts estimated 5 percent redundant parts. While implementing computerized parts management, it found by noun and description that 23 of 28 O-rings on a page of printout, each with a different part number, were the same. How did this happen? Each time an item of equipment was purchased, someone would order the recommended spares from the list in the back of the catalog. With a card-based system, there was no way other than experience to know the parts were already in stock. Further, the computer identified identical manufacturer part numbers. The true revelation came when the same parts were cross-referenced to equipment on which they were used and it was found that all the equipment had been removed the previous year. Computers can help save money!

FUNCTIONAL SPECIFICATIONS

The systems analysis process involves outputs inputs and process in that order, as outlined in Figure 16-1.

If you are already visualizing a computer terminal sitting on your desk or a microcomputer in the stockroom, stop right now. That is part of the last step in developing the service system. The first, and by far the most important, step is to determine exactly what is needed. A proven method is to assign a small group of user personnel, often called an "analysis and design team," to specify the system and be responsible through implementation, possibly a year or more later. A second management team should receive written and presentation reports and oversee the work, with check points and approvals at important phases. This helps provide high quality at minimum cost through Japanese-style group participation and ownership of results.

A good way to start this process is to schedule a brainstorming session of all involved persons who would use such a system. Have them prepare in advance a list of what they want the system to accomplish. At the meeting, list all the items on flip charts so that everyone can see them and the chart can be retained as a future check list. Do not be constrained by anything at this point. This is a "wish list" that will be prioritized later. An important part of this process is participation by everyone who may be affected. This human factor is vital because people are what make a system succeed or fail. Even the best designed and programmed computer system will not succeed without the active participation of those who use it. Some technically poor systems have been made successful by champions who really needed results and fought to make it work.

After the brainstorming session, produce minutes and a list of the requirements organized by functional purpose. The functional managers can then review the draft in detail with all their involved personnel and get everybody's ideas along with a feeling of participation. It is imperative that enough time be spent to identify needs and specify them adequately for systems ana-

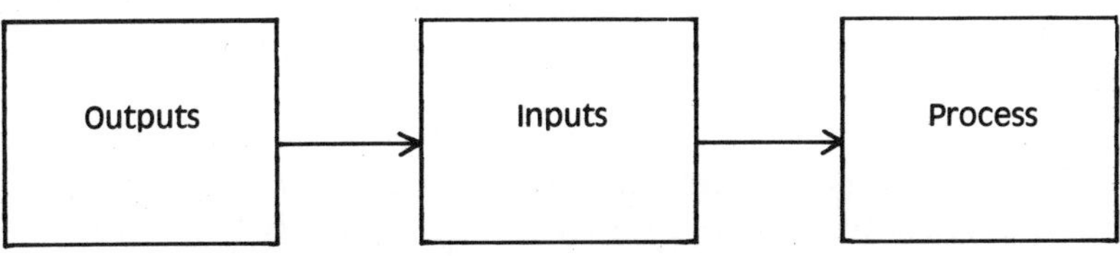

Figure 16-1   Systems Process

lyses. The tendency is for involved line managers to have an idea of what they want, but fail to convey it in adequate specifics to those who are going to create the system. Every data element necessary on an output screen or report should be defined by title and field length, whether it is an numeric or alphanumeric field, and whether it can be edited for specific entries. For example, if part numbers are six all-numeric digits, and are expected to remain so for the life of the system, then it should be specified that way. If there is any thought of changing to a longer field or adding an alphabetic requirement, that should be considered now because it will raise havoc at any later date. Like most projects, the best way is to do it right once.

Once the output specifications have been established on paper, the inputs should be determined. If existing reports are adequate, they may be used. If changes are required, the impact on personnel and procedures must be carefully evaluated. It is critical that input data be accurate. That is the origin of the saying, "Garbage in equals garbage out." The best way to assure accurate input data is to make it easy for people to provide the data in the normal course of activity, and then motivate them by showing improvements based on their input. Consider using bar codes, wand readers, mark sense, check blocks, optical character recognition (OCR), and even voice entry as means of improving input data.

Finally, after both the outputs and the inputs are determined, the processing middle step can be detailed. It should be only as complex as is necessary to process the inputs into timely, accurate outputs at minimum resource cost for time, people, and money. Functional specifications may be interpreted as best achieved by a card file, pocket notes, or word processing instead of the computer that is envisioned by many people. A system must flow logically from input through process to outputs. A good trial is to develop the system manually and make sure it works before putting it on computer. Anything that can be done well manually can usually be done well on the computer, but the investment is much greater so it should be tried manually first.

There are two approaches to developing systems. One is to spend considerable time and effort in detailing specifications that are then further analyzed for systems effectiveness, and then programmed, tested internally, and tried by the users. The other approach is to spend less time on specifications and get a skeleton system in place as quickly as possible so that users can try it and through actual hands-on experience determine what should be improved. There are advantages and disadvantages to both strategies. If the system is going to affect large numbers of people and will have extensive impact, then a large investment at the front end will pay off. The actual programming effort will usually be less that way. However, if time to implementation is important and the prospective users are not sure exactly what they want, it is best to get the skeleton system in place as soon as possible and let the users play with it. This is especially true with interactive, on-line computer systems where it is difficult to describe on paper the human factors of screen displays. Having used both extremes of careful specification and front-end planning versus rapid trials, this author has achieved best results on modern VDT-based interactive systems with the "quick and dirty, then clean it up" strategy.

USER-FRIENDLY GUIDELINES

Table 16-1 outlines requirements that have been gained from extensive experience with VDT-based service support systems.

BATCH VERSUS INTERACTIVE

The similarities and differences between batch and interactive computer processes are much like those between batch and process manufacturing. Batch systems operate by combining all the ingredients—data and computer programs—and running them as a single defined effort. Process manufacturing, however, like interactive computer systems, may be continually adjusted, with every change immediately altering the process. Both procedures are used in service parts management systems.

Development of cathode-ray tubes (CRTs), also called video display tubes (VDTs), and monitors with attached keyboards has allowed immediate display of information as it is processed. Data can also be continually updated. Batch reports may be displayed on a VDT screen, but are more commonly produced on paper printouts. A printout is historical information the instant the computer provides the information to the printer. Printouts are, however, useful for information such as parts catalogs by name

Table 16-1   VDT Guidelines

1. Arrange information in human thought sequence.
2. Put the most commonly used information first.
3. Use menu order and hierarchy.
4. Answer one question or objective per screen.
5. Provide logical access to following process steps.
6. Enable direct movement from any program to any other, regardless of menu hierarchy.
7. Edit all possible entries against tables, numerics, and logical dates.
8. Verify all possible coded entries with an immediate displayed description.
9. Block entries for transmission speed, consistent with immediate edit.
10. Allow search by name where identification code may not be remembered.
11. Use standard abbreviations.
12. Set half-density (dim) headings with full-intensity data.
13. Use alpha lock to make all entries in capital letters.
14. Size fields for standard conventions.
15. Provide cursor advance when field is filled, with New Line option.
16. Allow easy correction of user mistakes.
17. Suppress leading zero (0) entries on display and print.
18. Allow flexible date windows.
19. Establish logical hierarchies from the top down.
20. List entry options on the screen, when few.
21. Display all sections to be made on the screen at one time rather than one at a time.
22. Indicate any actions taking place within the computer, if more than 3 seconds are required to complete them.
23. Sort displays for user convenience.
24. Display specific messages that direct users to solutions.
25. Show the line last entered where multiple line scroll entries are possible.
26. Use predefined function keys on smart and intelligent terminals to speed user functions.
27. Provide print options via CRT, but only if paper is vital.
28. Print any paper reports on 8½- by 11-inch plain paper.
29. Allow traveling managers access to information from whatever CRT they may use through user name and password security.
30. Soft code as many variables as possible, and use parameter-driven values for flexibility and reduced maintenance cost.
31. Permit human users to make the final decisions.

or number that may contain thousands of items that are frequently referred to, but change little during a month.

Interactive computer processes consume computer capability on a high-priority basis. Frequently many users on an overloaded system will observe the system response time as too slow. If only a few users are active, the response time from one program to another, or from newly entered data to calculated result, may be microseconds. As the number of users increases, that delay between action and reaction will increase and could go into minutes. That is frustrating! Most interactive computers operate on a prioritized polling system that rapidly queries each user and either accepts some of the information the users are providing or produces for them some of the data they have requested. The answer to making this process perform faster is either more computer memory and input/output, or possibly redesign of the software programs that instruct the computer how to process the information.

To sort through 20,000 parts to find specific information such as bin location, and arrange the parts in that order, may require an hour of central processing unit (CPU) time, and often another hour to print it. To tie up a computer during mid-day to run such a report interactively is not wise use of the capability. The alternative is to run those same programs in batch mode overnight when few other users need fast response times. Many computer operators start their batch programs when they leave in the evening and let the printer run so that the printouts will be available the next morning for the time-consuming requests. Some systems can be programmed to query memory electronically at a specific time and perform any batch operations that are waiting in the queue. Note that information processed that way should be printed out. It is possible to display interactive immediate information on printout as well as on the VDT screen, just as batch reports can be prepared in either way. It is most useful, however, to have batch reports produced on paper.

ELECTRONIC IMAGES VERSUS PAPER

There are many advantages to having electronic images displayed on a screen. The information is instantly available and avoids the supply and disposal problems of consumable paper. Also, VDTs are much more reliable than printers. On the other hand, there are advantages to paper printouts. People like paper, are used to it, and can write on it; paper has an aura of legitimacy and permanency. A few sheets of paper are much less bulky to carry than a VDT screen. Paper is the only effective choice if a report has to be evaluated while one is on an airplane or in a motel room or other facility that does not have a terminal available.

Printouts are also advantageous for long listings such as parts catalogs. It is very helpful to have printed parts catalogs arranged by both part number and noun description so that personnel can review them quickly and find the parts they need. A catalog by vendor part number helps some organizations. Also, computers do fail sometimes, and must be down for preventive maintenance as well, so a paper catalog provides backup when instantaneous electronics may not be functioning. Even with reports that display variances only and the most important first, going through many screens to find the desired information can be more time consuming than looking at paper for the same information. The human eye can scan a full sheet of information quickly, so changing from one screen to another wastes time if the response is long. This is the same principle that makes displays of complete screens of information faster to use than line-by-line displays.

The best size for a paper printout is $8\frac{1}{2}$ by 11 inches or the comparable metric 21- by 28-cm A4 size, preferably on plain paper. The $8\frac{1}{2}$- by 11-inch size has two advantages: First, it is the same size as standard copies, notebooks, clipboards, and files. It fits easily into a briefcase and is more convenient to work with than the 132-column-wide paper often used. Second, the size emulates the VDT screen size. An 80-column-wide VDT image printed ten characters to the inch is 8 inches wide. That fits neatly on $8\frac{1}{2}$-inch-wide paper. The VDT screens that print six lines to an inch are usually 24 usuable characters high. Some printers have eight lines per inch and other options, but six is typewriter standard and is easiest to read. Of course, VDT images scroll vertically, so several screens can be printed on one sheet of paper. While a few reports may be closely spaced in 80 columns as compared with 132, the waste normally found on the larger paper is excessive. To date the author has not found any report that could not be adequately displayed in 80 columns. If more than 80 columns are required, then compressed printing to fit 132 characters into 8 inches is possible. The advantages and usefulness to human managers overcome any disadvantages.

It is an anomaly that capabilities exist for an all-electronic paperless system, but people insist on paper. The demand for printouts is actually greater when a new service part system is implemented than it is once the system is functioning well. People have to become confident that the CRT display will be accurate and available whenever they want it. If personnel are used to seeing reports in a particular form, it may be advisable to replicate that information initially, and also meet special requests in this way initially to help people feel comfortable with the new computer system.

HARDWARE

Hardware, meaning the durable, visible machines, are generally the items illustrated in Figure 16-2.

The central processing unit (CPU) is the brain that performs manipulations such as arithmetic addition and multiplication on the numbers and sorts them into desired order. One measure of the power and size of a CPU is the amount of main memory. The measurements are stated in bytes with the prefix kilo (k) meaning thousand, or mega (M) meaning million. A small personal computer would typically have 64 kilobytes (kb), which means its main memory can work on 64,000 bytes of information at a time. Each byte is roughly equivalent to one letter, or number, or symbol character. Larger computers have memory measured in megabytes (Mb).

The size distinction between micro-, mini-, supermini-, and mainframe computers is blurred. Similar microelectronic technology is present in all sizes. There is simply more of it arranged in different fashions in larger equipment. Current technology can place 64 kb, 256 kb, or even 1 Mb of memory on an electronic IC about the size of a fingernail.

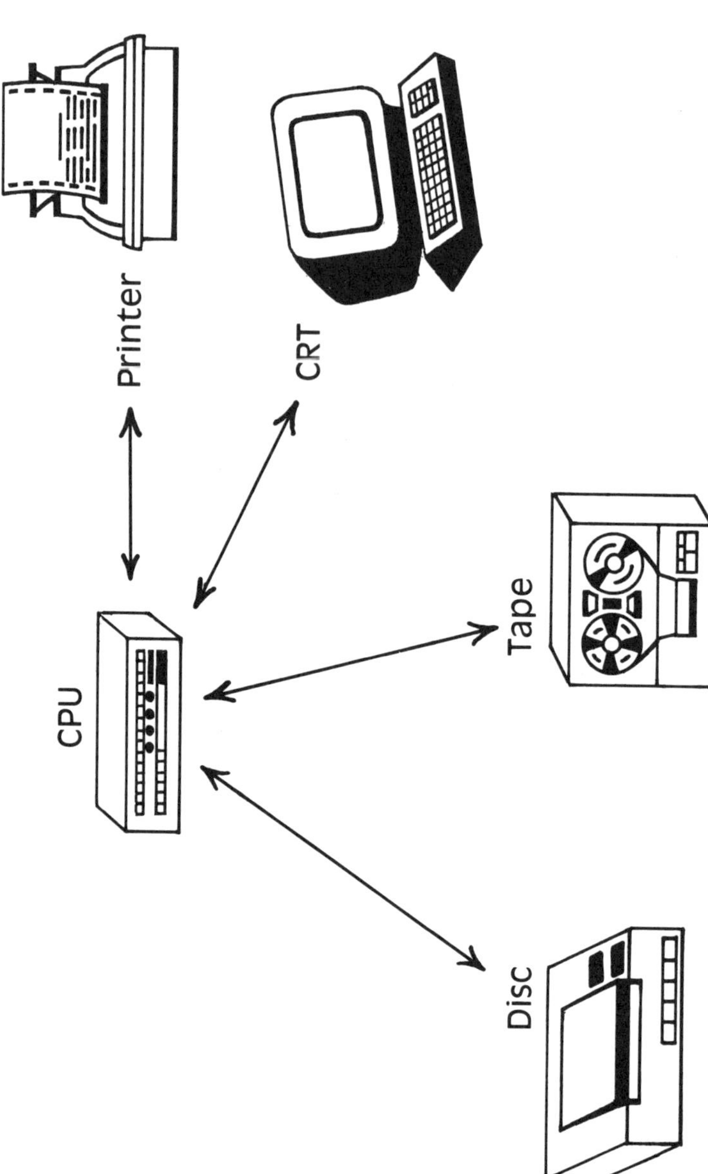

Figure 16-2 Basic Hardware Configuration

An entire microcomputer CPU can be located under the keyboard. Even large minicomputers with memory to 12 Mb can have the entire electronic package located in about 3 cubic feet. Superminicomputers today are usually 32-bit machines capable of performing what only large mainframes could do just a few years ago. IBM reports, for example, that in 1955 it would cost $7 and 14 minutes to compute information that today can be calculated for $1 in 1 minute.

Computers operate on digital electronic memory. Each bit is either charged or not charged. It is like a light switch that is turned either on or off. Eight bits together can form the code structure for 1 byte. This multiple of eight is why computers normally are structured as 8 bit, 16 bit, 32 bit, and so on, although there are 12-bit machines and other variations for special purposes. The more bits that can be transmitted at a time, the faster and more efficient the computer can process information. The type of memory used in the central processing unit is normally called RAM, which stands for random access memory. It can be recorded, played back, written over, or erased any number of times. When electric power is removed from RAM, all memory is lost. Some computers have a battery backup system to prevent this from happening until main power can be restored. Read-only memory (ROM) is used for information such as operating software that must be retained and not erased or written over. Special kinds of ROM such as PROM (programmable read-only memory) can be reprogrammed in the field for specific functions.

Since main memory on a computer is relatively expensive, disks and tapes are used as peripheral storage media to retain large amounts of information. Tapes are least expensive. Microcomputers often can use a standard audio tape cassette to load information. Larger computers operate on larger sized tape that can retain much more information and run at higher speeds. Tape is very efficient for storing sequential information such as parts lists that is always wanted in the same sequence. Like an audio tape, to find a specific selection may require fast forward through the entire reel of tape and then searching to find the specific item of data. For this reason tapes today are used primarily to back up the information that is usually kept on magnetic disks.

Magnetic disks range in size from 3¼-inch pocket-sized floppy diskettes to 5¼-inch floppy diskettes that are commonly used on microcomputers to large fixed Winchester hard disks that can hold over 600 Mb of data. A 5¼-inch floppy disk looks much like a 45-rpm record in its cover. The larger disks are more like stacks of 33-rpm phonograph records separated by enough space for the read/write heads of the disk drive. A magnetic disk, like a phonograph record, allows any information at any point on the disk to be found quickly by moving the head. The disk spins at high speed and the head moves in and out. A disk is electronically formatted into concentric tracks and then pie-shaped sectors that radiate as wedges from the center out. The combination of track and sector provides an address for data that can be rapidly located by the computer's electronic memory map. A disk must be formatted to the specific model of disk drive, and sometimes even to the specific drive in which it is to operate. Floppy disks from one brand of personal computer, for example, may not operate in another brand because the electromechanical characteristics of the drives are different. It is possible in most microcomputers to trade disks, for example, from one IBM personal computer to another, but not to a Radio Shack or Apple computer. Disks may be recorded, erased, and rerecorded many times. It is possible to write-protect large disks electronically. Floppy disks normally have a notch in the cover edge that can be taped over to protect valuable information.

The combination of on-line disk memory and main RAM memory is important in the use of a computer system for managing service parts. The larger the memory, the more information that can be kept on it at any time. On a 360-kb floppy diskette, efficient software can handle about 2000 parts, with each part record containing about 150 data elements. If more parts than that are to be managed, or more information is necessary on each part, then more diskettes must be used.

Input, in addition to a keyboard, may include mark-sense readers, optical-character readers, bar-code wands, touch-panel VDTs, and voice recognition. The fact that relatively few service parts personnel can type well inhibits entry and retrieval of data. Since keyboards are the prime medium today, accuracy in reading and entering long strings of numbers and letters is often deficient. Therefore, emphasis is being put on devices to enter parts information accurately and rapidly into computer memory. Bar codes are useful in high-volume operations. The U.S. military now requires them for identification of parts and equipment. The most useful bar-code reader for the parts business is attached to a portable recorder. The part identification is indicated on the bar code, as is probably the bin-location identifier. After the bar code has been scanned with the wand, quantity counts can be entered via the numeric key or symbol pad. Allowing that same information presently entered on the keypad to be voice entered will be a further advancement. Voice-recognition electronics is available today that can be voice patterned to recognize the individual user's voice, and to allow entry verbally of typically 200 words. That is adequate for the parts business. The transcription device translates the voice into digitized alphanumeric characters. It is highly accurate on numbers and letters. As such devices have already been demonstrated for complete word processing, it will obviously, in the future, allow complete data entry and command retrievals for computerized parts management.

Optical-character readers (OCR) are also becoming more sophisticated, more accurate, and lower in cost. Sophistication is necessary because of the variation in people's handwriting. This technology has been used for several years on service call reports that are carefully printed. More latitude in recognition is available today, and prices are getting low enough that OCRs are becoming more practical. Light pens and touch panel screens are useful for selecting the particular element desired on a screen display, but have limited usefulness for parts.

Printers for parts management come in three basic varieties: letter quality, dot matrix, and high speed. Those terms are a mix of technology and capability. Letter-quality printers usually have a formed character on a daisy wheel or thimble that is struck by a small electronic-driven piston hammer to put the ribbon ink image accurately on a printed page. These may be necessary for high-quality correspondence to customers. They are generally slow speed, in the area of 30–55 characters per second (cps), and can cost several thousand dollars when equipped with sheet feeders that load an individual piece of paper at a time. Dot-matrix printers operate on the principle of small wires that are arranged in the matrices, typically about seven by nine, being activated to strike the ribbon and produce an image on the paper in the same pattern as the desired letter or character. The speeds of dot-matrix printers typically range from about 80 to 300 cps. Enhanced images described as correspondence quality are possible on some dot-matrix printers by making a second, or even third, pass over the image, often slightly offset to fill in the spaces between the original dots. Naturally printers operating in the enhanced mode cannot produce as many copies per second as they can at slower speeds. To be more versatile, graphics capability is often included in their features. High-speed line printers operating with line, band, drum, or even laser technology are useful for printing long lists such as parts catalogs. Their speeds are rated in lines per minute (lpm), so that a 300-lpm printer is equivalent to five lines per second, or about 300 cps. The actual throughput speed will usually be something less than any printer's ratings. Most are microprocessor controlled to skip blank spaces and print backward as well as forward for efficiency and less stress on the print-head return. Line printers are available in speeds to 2000 lpm, although the price increases rapidly beyond 600 lpm. Good-quality fanfold, tractor feed paper is the only way to go at high speed, and is beneficial even on dot-matrix printers. Colored plotters are available for graphics, but are of little value for service parts management. Much numeric data can be displayed better in graphical form from plotters or on bar charts from regular image printers.

The VDTs come in three main levels of ability: dumb, smart, and intelligent. A "dumb" VDT typically has the same keys as a typewriter keyboard, generally with a numeric pad. Such terminals are of limited use and are fading in popularity as more advanced features become available at reasonable cost. "Smart" terminals add function keys that can be preprogrammed to provide backfield, end input, scroll, change program, help, and other useful movements. "Intelligent" terminals have additional memory and are essentially microcomputers themselves that can process information locally and then communicate with larger minicomputers or mainframe computers.

SOFTWARE

The computer code programs and other instructions that tell a computer what to do are called software because they cannot be seen as hard objects can and yet are critical elements of providing a computerized service parts management capability. The operating system consists of instructions that tell the computer how to get information from inputs and memory devices, how to process it, and how to communicate it properly to memory and output devices. Names such as CP/M, PC DOS, and Unix indicate common operating systems for microcomputers. The fact that there are some common systems greatly facilitates exchange of software between computers. On minicomputers and mainframe computers names such as AOS and VMS include the initial S, standing for operating system. A letter V typically stands for virtual. Understanding of operating systems should be left to computer experts, but service parts managers should realize that they are a necessary part of a computer's capability.

Database management systems arrange the data in file structures, usually sequential, random, or relational, that store and retrieve the data efficiently within the computer's memory. Some newer databases on microcomputers are actually languages themselves, as well as data managers.

Languages are the forms of code used by a human to instruct the computer. Typical languages include:

ADA—The new U.S. government standard.

APL—"A Programming language" commonly found on time-sharing networks. Efficient for short engineering calculations.

Basic—"Beginning All-Purpose Symbolic Instruction Code" is the initial language of most microcomputers, but unfortunately is not well standardized.

COBOL—"Common Business Oriented Language" is most useful for finance and maintenance and service-type systems where many numbers have relatively little done to them but are mainly added, subtracted, divided, and multiplied.

dBASE II and III—popular microcomputer relational database and program language © Ashton-Tate.

FORTRAN—"Formula Translation" is the standard language for calculations that require extensive manipulation of numbers.

There are many other languages such as C, Dibol, Pascal, and RPG that are either special purpose or machine specific. For service parts management, COBOL is the best for large computers since more programmers know it than any other language, and it is well structured, relatively easy to maintain, and efficient. On microcomputers, experience has shown that dBASE II and III are faster to program and more efficient than either COBOL or BASIC. Programs written in a language such as COBOL that can be interpreted by a human are called source code. Within the computer that source code is compiled into object code that is readable only by the machine. Most programs are sold or licensed as object or source code so that the users cannot change them. Software provided in object or source code form is typically warranted against all defects for one year, but all warranties are voided if the source code is altered.

Computer power gives service parts managers a major new tool. The challenge now is to learn to use the technology to help humans manage the parts business.

17

Personnel
and
Organization

JUST IN CASE THE POINT has not already become clear, it is emphasized again that service parts management is a people business. All the technology of computers, sophisticated equipment, and models cannot accomplish much if good people are not involved. The abilities and motivation of stockkeeping personnel is of vital importance. Good people can overcome any problem. Every stockroom, no matter how small, should have someone responsible for its operation, with authority to do whatever is necessary to provide good customer support.

Successful organizations have realized the synergistic influence of proper parts management in being able to multiply the effectiveness of other service personnel. The manager should have experience with parts; understand statistics, risk taking, and supervision; be highly motivated; know where different kinds of parts are used, and what the necessary storage conditions for them are; and be able to manage any subsidiary personnel. Pickers and packers may be new personnel in an organization. This can be an entry-level position for people who can read and write and pay attention to detail. One of the most imprortant characteristics of a parts person is attention to quality that will assure picking the proper part and handling it safely so that the correct item arrives at the user's location in good condition.

QUALIFICATIONS

The qualifications listed should serve as check lists.

Stockkeeper Qualifications

1. Education—high school or equivalent.
2. Experience—helpful, but not mandatory, to have worked with equipment using parts similar to those supported.
3. Special qualifications:
   Ability to read and write accurately and neatly.
   Ability to enter data on keyboard terminal with at least two fingers at a time.
   Ability with numbers generally equivalent to balancing a checkbook.
   Thoroughness.
   Consistency.
   Honesty.
4. Responsibilities:
   Establish and care for parts-keeping facilities.
   Maintain accurate parts records.
   Determine when parts need to be reordered.
   Obtain parts to meet special needs.
   Expedite parts for emergency repairs.
   Obtain and support consumables and supplies purchased in bulk but issued in smaller quantities.
   Investigate and recommend vendors who can meet supply criteria.
   Fill out purchase requests.
   Receive parts from shippers or plant receiving and assure that proper items and quantities are received and stock them to the proper locations.
   Issue parts to authorized requestors.
   Receive unused parts back into stock.
   Arrange transportation for parts as required.
   Return repairable parts to center; assure requested data are provided and that credits are accounted for.

Inventory Analyst Qualifications

Large support operations with over $500,000 in inventory should utilize the services of an inventory analyst. The analyst should be concerned with forward planning, including provisioning for new products and establishing recommended spare lists. He or she should evaluate the parts in inventory, analyze both excesses and shortages, and act in coordination with the stockkeeper as necessary to bring stocks into balance.

A typical job description for an inventory analyst is as follows:

1. Education—a bachelor's degree or the equivalent in business, engineering, mathematics, or a scientific field with emphasis on numerical analysis.

2. Experience—at least two years' related experience such as production, inventory control, or field service management, or with a parts distributor or retailer.

3. Special qualifications—hands-on experience, which should provide knowledge of what kind of parts are potential problems; ability to relate part names to the physical objects and the conditions under which they are used.

4. Responsibilities:

In cooperation with technical personnel, analyze new products to determine which parts will need to be replaced.

Determine provisioning strategy including where stocks should be located and quantities at each location.

Establish initial economic order quantity, and request parts through purchasing channels.

Determine initial reorder point.

Coordinate with engineering and marketing to assure that inventories are adjusted for revisions to parts, special projects, end-of-life buys, and other changes.

Provide input to life-cycle costing and trade-off analyses that involve parts.

Provide ongoing measurements to determine when parts are above or below control limits.

Prepare budgets and operate within approved limits.

Parts Procurement Qualifications

Parts procurement in a small organization may be embodied in the same person who does the inventory analysis, and probably brews the coffee too. In larger organizations purchasing may be a completely separate function to which the inventory management simply forwards purchase requests. Responsibilities of the part's purchasing function, regardless of where it is located, include:

To develop and keep current a list of authorized vendors.

To track vendors performance to determine the recommended vendor for each part.

To process purchase requests for necessary approvals.

To create purchase orders from the approved requests.

To consolidate purchase orders to get the best possible combination of vendor performance and benefit/cost.

To telephone prospective vendors to determine whether parts are available and to request quotes.

To prepare formal requests for proposals or quotes, where required.

To negotiate any discrepancies, such as shortages or quality rejects.

To assure tracking of receipts to accounts payable and authorization of payment.

TRAINING

Formal training with on-the-job follow-up is vital for qualified parts personnel. As many procedures as possible should be developed to show the "one best way." They should be written and used as a core for the training program. It is easy for someone long experienced in the parts business to say that things are "intuitively obvious and anybody can do that." Quite often that is not true.

When a person is assigned to the stockroom, an initial orientation walkthrough should be provided by the supervisor. Then a classroom session should be conducted to outline all the activities that go on and how the procedures meet those demands. Forms should be completed during training to assure that all entries are complete and legible. If a computer terminal is used, the available programs should be exercised with training that will point out any errors. Self-paced instruction is possible for much of this training. If the turnover in personnel requires frequent training, then an investment in programmed instruction, either written or on a computer, will be helpful.

After formal training is completed, the new person should be assigned a mentor. This experienced person should offer assistance when required and keep track of the new person's performance to assure that everything is going well. Most experienced persons will take great pride in this opportunity and responsibility. Watch out, however, for the temptation to conduct all training on-the-job by the old employees. Make sure the way things are being done is the way they should be done, and that high-quality instruction is conducted in a professional manner. There is a saying in the training profession, "You pay for training whether you need it or not." That means that if you pay for training initially, performance will benefit. If training is not provided, then performance will be poor and the cost will be imposed as a penalty on low productivity. Good training in service parts management is a worthwhile investment.

MOTIVATION

Motivating parts personnel is not an easy job. It often seems that as soon as one challenge is met, several more crop up in its place. Helpful factors include:

1. Management attention
2. Objectives, goals, and measurements
3. Good working conditions
4. Solicit participation
5. Delegate
6. Provide enrichment
7. Allow authority

The effect of management attention has been well known since the Western Electric Hawthorn studies showed that telephone assembly persons responded more favorably to the personal attention they received than to lighting, colors, and other physical parameters. It is often noticed that maintenance and production personnel who feel that no one cares what they are doing will cause a problem just to receive attention. That is a childish trait that is carried through to adulthood. Parts personnel are often neglected. They are criticized when wanted parts are not available. They get little praise when things go well. Management needs to spend more time on helping parts personnel know when they are doing well or badly, and publicly praising them when appropriate and providing constructive criticism when necessary.

One of the best techniques for motivating stockroom personnel is a large wall chart that tracks number of orders picked, packed, and shipped, percentage on time, percentage error-free, and so forth. A typical chart in reduced form is shown in Figure 17-1.

PARTS PERFORMANCE for 3/7

Orders requested	530
Orders shipped	525
Number of line items	1575
Percent on time	99
Percent error-free	99.8
Percent demands filled	97
Number of complaints	0

Figure 17-1 Parts Performance Motivational Chart

Current data can be entered on the chart at the mid-day break and again at the end of shift. The previous day's data would be left until noon of the next day. Some things are beyond the control of stock personnel, such as the number of requests. You do not usually pack orders that have not been requested. But on high order days, the orders shipped becomes a goal to strive for and helps motivate personnel to work faster to get the new record on the board.

Soliciting ideas from the workers is another very effective technique. Many managers are surprised at how much employees really have to offer when the manager opens up enough to request their help. Suggestion boxes are useful in a large organization, but the most effective approach is from the supervisor who says, "We ought to be able to do this better. Do you have any ideas?" Good follow-up is necessary, of course, and the ideas should be put to use whenever possible. Soliciting and then not doing anything will put an end to ideas. Publicly surrounding such ideas usually brings out more. Participation groups such as the quality circle are also effective. Employee groups can address specific problems or may address broad issues that can then be narrowed down into specific challenges. Participative approaches by a facilitating leader can go further toward good parts management than any amount of dictatorial demanding.

The management techniques of providing authority commensurate with responsibility and delegation are also effective. A human tendency is to do something yourself because it is quicker and more accurate than is taking the time to explain it to someone else. The result for employees can be a very dull work situation. It is much more exciting if their jobs are enriched by opportunities to do special projects and stretch their talents. As many tasks as possible should be delegated to the lowest level person who has the ability and knowledge to accomplish them.

Authority, responsibility, and accountability should go hand in hand. A manager can delegate authority but will not be relieved of the ultimate responsibility to assure that the job is accomplished. The person who is given and accepts the task is made responsible and accountable for accomplishing what that person has agreed to do. There are many difficult situations in managing parts where someone is responsible for something without having the necessary authority. For example, a person is told to do everything necessary to get a job done, but then is restricted as to the personnel and resources available. Or the parts manager is held responsible for a 98 percent stock level but is not given authority to determine which parts are to be ordered and how many dollars can be spent. If a job is to be accomplished, then authority must be given that is commensurate with responsibility. The complete combination of authority, responsibility, and accountability will help provide strong motivation for parts personnel.

18

Parts Improvement Programs

VERY SERVICE ORGANIZA-
tion can be improved. Continuing changes in technology, equipment, customers, personnel, economics, government regulations, competition, and management interact to affect performance. Chances are that your parts organization is not even sure how good it is today. Are your service representatives carrying the right parts? Or is excessive time spent in parts chasing? How much money is invested in parts?

In most companies top management recognizes that service is critical to the company's success. The marketing vice president needs increased sales to satisfy customers, the controller wants expenses contained, and the president looks straight to the bottom-line profits. Every organization, large and small, should step back frequently, look in the mirror, and conduct a parts improvement program (PIP).

The objective of evaluating field service performance is to measure quantifiably how well an organization is doing today for comparison against objectives, and determine corrective action where necessary. It should establish facts that can place a priority and dollar value on each item. A parts improvement program can be included in a larger service improvement program that includes all aspects of operations, training, tools, technical support, administration, and parts.

Is this an audit? Yes, but it goes far beyond an audit because its objectives are not only to gather facts and measure them, but also to develop practical solutions to identified problems and opportunities, and assist in the implementation of the improvement actions. Experience has shown that most service organizations do not obtain adequate results from just an audit. The report too often is dropped, persons who should be improving feel that the identified problems certainly could not be happening in their area, and the desired improvement change generally does not take place. A PIP is intended to carry the efforts on to practical implementation.

STEPS IN A PARTS IMPROVEMENT PROGRAM

The major steps are as follows:

1. Establish objectives and goals.
2. Determine what measurements are necessary to measure performance both quantitatively and qualitatively against the goals.
3. Determine the inputs necessary to gain those output measurements.
4. Develop the measuring forms and other instruments necessary to gather the input facts.
5. Detail the process that will be used to convert input data into output answers.
6. Determine the people, financial, and other resources necessary to accomplish the project.
7. Integrate these elements into a coordinated plan.
8. Gather the data at headquarters and in the field.
9. Analyze the results and develop corrective actions.
10. Present the information to management.
11. Develop corrective actions for representative situations.
12. Implement the corrective actions in representative situations.
13. Revise as necessary.
14. Implement uniform corrective action everywhere possible.

These steps are illustrated in Figure 18-1.

PLANNING

Thorough planning is vital! As the old saying goes, "It is not easy to get there if you don't know where you're going." Therefore, objectives and detailed goals must be established by top management as guidance for the plans.

Typical of information seeking, outputs are the first thing to be established and then the inputs and processing are determined. Many of the outputs will be in the form of questions to be answered, such as, "What is our current productivity?" Everyone concerned or who might benefit from the improvement project should be asked for their needs and suggestions. After all the desired outputs are gathered, grouped, and prioritized, the facts necessary to gain those measurements must be determined. A planning form helps to ensure that every question can be answered and is further linked to a specific data-gathering form and analysis process.

Figure 18-2 shows a typical form used for recording activities and times of field technical representatives. Figure 18-3 is a partial coding sheet that allows each activity to be recorded in logical sequence to the nearest minute and then analyzed us-

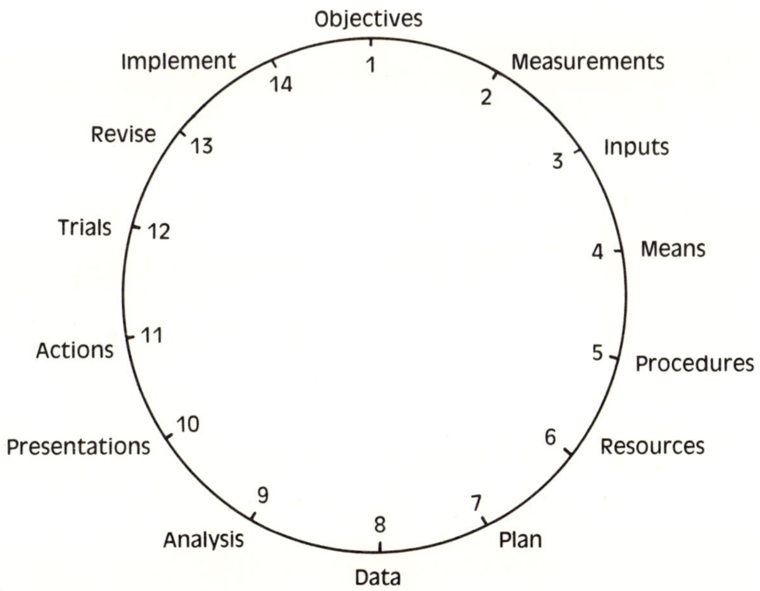

Figure 18-1   Service Improvement Program Steps

Tech Rep Name ___________________________   Page _______ Of ________

Observer   ___________________________

Location |_|_|_| Tech Rep Num |_|_|_|_|_| Date |_|_|_|_|_|_|
                                                             MM DD YY

| Time Start | Act Code | Prod Line | Job | C/T/B | | Time Start | Act Code | Prod Line | Job | C/T/B |

Figure 18-2   Tech Rep Time Log

1 Branch office
  1A Paper work
    1A1 Mail
    1A2 Service call report
    1A3 Account Analysis
    1A4 Expense report
    1A5 Correct mistakes
    1A6 Special projects
  1B Parts
    1B1 Parts requisition
    1B2 Return to stock
    1B3 Obtain part number
    1B4 Pricing
    1B5 Travel to get part
    1B6 Checking on orders
    1B7 Canablization (total approval
      —removal
        of part—travel both ways)
    1B8 Defective return
  1C Tools
    1C1 Obtain special tool
  1D Telephone calls
    1D1 Customer assistance
    1D2 Assisting other TRs
    1D3 Tech support
    1D4 Other districts
    1D5 To schedule PM
    1D6 Other
  1E Meetings
    1E1 Central branch
    1E2 Service Department
    1E3 Group
    1E4 Manager/supervisor
    1E5 Check with dispatch
2 Travel
  2A Office to customer
  2B Customer to customer
  2C Customer to office
  2D Gas and other car maintenance
3 Nonproductive
  3A Break
  3B Meals
  3C Check parts locker or mail
  3D Personal
  3E Search for customer location
4 Customer's office
  4A Entry/access to equipment
  4B Diagnostics
    4B1 Using diagnostic tools
    4B2 Board swapping
    4B3 Service literature
    4B4 Disassembly to inspection
    4B5 Observation
  4C Repair
    4C1 Adjustment
    4C2 Part replacement
    4C3 Clean or replenish
    4C4 Temporary repair, need parts
    4C5 Reassembly and checkout
    4C6 TR personal cleanings
  4D Customer training
    4D1 Machine operation
    4D2 Operator functions
    4D3 Meet with customer
  4E Preventive maintenance
    4E1 Contract inspection
    4E2 Planned PM
    4E3 Convert CM to PM
  4FPhone call for tech asst
    4F1 To supervisor
    4F2 To other tech rep
    4F3 To region specialist
    4F4 To home office
    4F5 From other tech rep
    4F6 From dispatch
  4G Telephone calls for parts
    4G1 To supervisor
    4G2 To other tech rep
    4G3 To stockroom
    4G4 To distribution center
  4H Telephone to dispatch,
    next call
  4I Retrofits
  4J Sales
    4J1 Canvassing
    4J2 Sales assist
    4J3 Demo floor equipment
  4K Shop work
    4K1 Repairing machine
      assemblies
    4K2 Bench work
  4L Paperwork
    4L1 Service tickets
    4L2 Other
5 Training
  5A Self
  5B Other tech rep
6 Installation
  6A Preinstallation
    6A1 Assure site ready
  6B Normal installation
    6B1 Uncrate/inspect
    6B2 Assemble
    6B3 Test, checkout
    6B4 Operating instruction
      —charge sales
    6B5 Operating instruction
      —no change
  6C Problems at installation
    6C1 Move equipment
    6C2 Clear area
    6C3 Diagnose problem
    6C4 Remove part
    6C5 Adjust, repair
    6C6 Get part number
    6C7 Call for part
    6C8 travel to get part—
      Wait for part
    6C9 Replace Part
    7A1 Other ________________
    7A2 ____________________
    7A3 ____________________
8 Deinstal
9 End of time study

Figure 18-3   Service Time Study Codes

ing computer programs. Note the part activity codes in 1B, 3C, 4G, and 6C.

SELECTING THE SAMPLE

It is generally necessary to select a sample of organizations and activities rather than to try to evaluate all elements of the total population. Main considerations are as follows:

1. Representation
2. Influence
3. Inclusion of extremes of geography, products, customers, environment, and perceived performance
4. Dispersal
5. Image
6. Cooperation expected
7. Size
8. Convenience of access
9. Cost in money and time

If an organization is divided by geography, it is a good idea to ensure that a suborganization in every area is evaluated. This helps to consider climates, customers, environment, and other differences that may exist. It also helps to eliminate the statement, "It may be like that in their situations, but it couldn't be that way in mine." In fact, regional differences have been found to be very small. In one large organization, the productivity of districts centered in Columbus, Ohio; Detroit, Mich.; Miami, Fl.; Mountainside, N.J.: and San Francisco, Calif. were all within 1.5 percent of each other, although there were variances within the activities that make up productivity.

It is a good idea to select for evaluation the location that is generally considered the best, one that is midrange, and one of the worst. The reasons behind the differences are often useful in stimulating improvements. A small, a medium, and a large location should also be evaluated since there will be differences in their methods and problems.

Every organization has certain activities that are most influential and are highly visible. Major payoff usually occurs by starting improvement projects in an influential location so that word spreads quickly from an internal authoritative source that is "one of us" and establishes an image of the parts improvement program as a positive project rather than a "witch hunt." Obviously the planning must be done thoroughly so that you walk into the first location well prepared.

TYPICAL PARTS IMPROVEMENT PROGRAM TEAM

Notice that the word *team* was used instead of just personnel or people. That is done because the team must be made up of top-quality experts from all necessary functions. Figure 18-4 shows a typical team. The team could be provided by external consultants. However, best results are obtained through maximum use of internal personnel, with consultant assistance.

Recruiting the team members must have top management support. A good procedure is to require every field region to

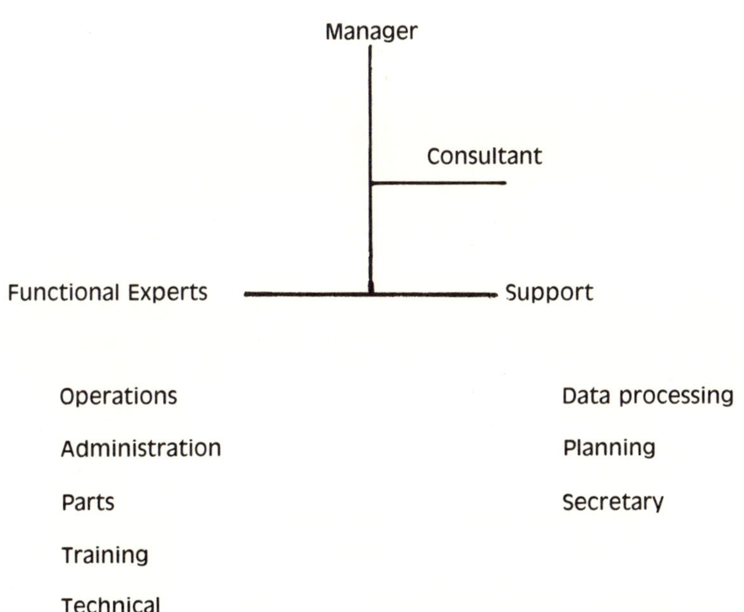

Figure 18-4   Improvement Team

offer the names of two top managers so that a field of perhaps a dozen candidates is available from which one to three would be selected. It may be necessary to request personnel specifically by name from headquarters staff.

Participation in a parts improvement program is a major developmental experience that cannot be gained any other way. If the improvement program is going to require the individuals' full time for more than six weeks, they should be replaced in their existing positions. They should be promised in writing that, upon successful performance on the program team, they will be placed at the head of the candidate list for promotion and new assignments after the program is over.

On long tours, morale and family cooperation will be strained. A complete review of a company's field operations may require six to nine months, with full implementation stretching beyond that. Leaving one's family on an airplane every Sunday evening and arriving home again late Friday night from an exhausting week can quickly wear down even the best person. Team members should occasionally be able to get home early, and perhaps even bring the spouse to the field location for a weekend if it is a desirable spot.

Facilities at headquarters must also be provided so the team members have a place to sit, keep their files, and have secretarial support for typing, travel arrangements, and other concerns.

FACE-TO-FACE, ON-SITE EVALUATIONS

There is no substitute for fact finding and evaluation conducted where the action is. Perceptions may be very different from the facts as shown on a computer printout. Often the subjective, emotional information gained during an after-work meeting for a drink or coffee will identify the problem and suggest corrective action.

Remember that the end objective is to implement corrective action for the problems discovered. Many good ideas are just waiting for you to ask about them.

UNIQUE CONCERNS

When conducting a parts improvement program, keep your senses alert. It is easy to make premature judgments based on initial impressions. There is nothing as useless as solving the wrong problem, so make sure the facts are accurate.

Most good managers will welcome the improvement program with open arms, recognizing that they need all the help they can get. A few, however, will be defensive, usually because of their fear of being criticized. It can be very helpful to sit down with all affected managers and let them know in a business-like fashion at the outset why you are conducting the investigation and that the objectives are to improve service performance for everyone's benefit. The improvement program team manager must be a good salesperson with public relations abilities to sell the activities in a positive fashion, yet be backed up by top management that will dictate compliance if necessary.

Surprise is necessary to assure that realistic situations are being viewed. The informal grapevine in most service organizations operates with amazing speed. Only the improvement program team members and top management should know the organizations selected for evaluation. Activities most susceptible to bias should be evaluated before any change from the normal system can be induced. It is also possible to use control groups for comparisons. For example, when it was suspected that special attention for emergency parts was being given to a northern New Jersey branch being reviewed by a PIP team, the Philadelphia branch was asked simultaneously to submit orders for the same parts. No one was very surprised when the northern New Jersey branch received its parts within two days, but the Philadelphia group had not received its parts ten days later. Persons responsible for this bias were simply told that they had proven they could do the job well in New Jersey, so they had better get on with doing it as well everywhere.

Consultant assistance can be very valuable. Many organizations do not have the internal capabilities to do everything involved with an improvement program. Even when internal capabilities are available, an independent viewpoint and fresh pairs of eyes will identify problems and opportunities that may be otherwise overlooked by people who encounter them every day, and adding consultants to your evaluation team indicates to all concerned how important the improvement program is.

A service improvement program is expensive, but there is generally a payoff many times the investment. In one typical organization, the improvements from a complete program amounted to an increase of 33 percent more service calls per day at a reduction in response time of 50 percent and a 10 percent reduction in travel mileage. The benefit in just the first full year was about 100 times the cost. That is an effective ROI, and a sound reason to undertake a service improvement project.

19

Special Considerations

HOW MUCH INVENTORY IS "ENOUGH"?

THIS QUESTION HAS PROBably been causing ulcers since the first cave people worried about how much food and fuel would be necessary to get through the coming cold weather. Maintenance service personnel usually want more. Financial management pushes for less. And the best balance is probably somewhere between.

Beware of dogmatic, absolute rules such as "Inventory should be no more than 5 percent of sales" or "Inventory should never exceed $500 per service employee" or "Inventory must turn at least X times per year." Worthwhile comparisons can be made with other organizations in the same industry if consideration is given to geography, product mix, maintenance strategy, desired service level, and customer satisfaction. A dollar saved on inventory may cause a $100 loss of income. Having the right parts (quality) is more important than having a lot of parts (quantity). The real answer is found by gaining benefits from parts held that exceed the cost of holding those parts.

DEMAND VERSUS ISSUE

The distinction between demand and issue is important. Many parts are requested (demanded), but if they are not on hand, the requestor goes elsewhere and the demand is not registered. This can lead to a never-ending cycle of no stock = no issue = no stock. Demands should be recorded so that parts with sufficient demand are added to stock and those with no demand are eliminated. A good way to do this is to record the demand for a part and then check to see if it is available. If it is, then demand = issue. If it is not, then demand is registered and the part may be added for future issue.

STOCK MANAGEMENT

In a basic parts management system, the divisions of part types may be simplified into four:

Detailed stock
Summary stock
Nonstock
Insurance

Detailed stock means that the parts are used often and are of high value so that they should be accounted for by each individual transaction. Summary stock parts are of low value such as common hardware, and can be stocked in quantities of 50 pounds or by the coil or by whatever the measure is, and not counted individually. A good method of control is to have two boxes of parts. One contains the parts presently being used. When that box becomes empty, the reserve box is opened and at that time an order is placed for more of the parts.

Nonstock parts are not used frequently enough to justify keeping them in regular stock. The criteria for stock versus nonstock should be determined to meet needs at each stock level. Insurance parts are designated by management to be kept on hand just in case they are ever needed, even though they are rarely required. Highly essential, long-lead-time parts might be kept for insurance. It is possible to do a stock review based on about six months' history and evaluate the total stock for necessary adjustments. An example of printout messages from a stock management computer program is shown in Table 19-1.

Table 19-1   Stock Management Guidance

Part Number	Action/Recommendation
13224	"I" CODED ITEM HAS HAD 0 REQUESTS IN LAST YEAR. RECOMMEND STOCK CODE BE CHANGED TO "N" AND PART BE REMOVED FROM INVENTORY.
14134	ITEM NOT ON HAND, NOT ON ORDER, AND NO REQUESTS IN LAST YEAR. RECOMMEND DELETE FROM DATABASE.
14573	CHANGE FROM DETAIL TO SUMMARY. BAG QUANTITIES AT LOCATION BIN 12A AND ATTACH PHYSICAL REORDER CARD.
15245	SUMMARY ITEM REORDER POINT RECOMPUTED. CHANGE REORDER QUANTITY AT LOCATION A35D SHOULD BE 15.
15925	CHANGE FROM SUMMARY TO DETAIL. UNBAG QUANTITY LOCATION D35A AND DESTROY REORDER CARD.

The order-processing system generally should keep track of the "Date Last Issued" as an important data item. Experience shows that some parts will be demanded in the future even though there was no demand in the past 12 months. In several actual cases, 11 percent of parts with no demand in the past 12 months did show demand in the 13th to 24th months. Care must be exercised to assure that parts supporting equipment still in the field are not discarded just because there has been no recent demand. Some of the reasons for lapses in demand include:

1. High reliability of the part (which could begin to wear and fail).

2. Demand being absorbed by local or dealer stocks without replacement order.

3. Local "on-site" fixes that mask the fact of failure to the supply system (but which could quickly become demands if field policy changed).

IMPORTANCE

Extended cost, demand, value class, essentiality, control, and lead time have all been discussed as reasons to pay attention to specific parts. Experience has shown that only a few people can effectively consider all the individual factors when managing more than a few thousand parts.

A good resolution is to combine (homogenize) all the considerations into a single rating. Call it class and give each part a number 1 to 3 or letter A, B, C, with 1 or A the most important. Combine all the considerations into a single class for each SKU. For example, a main drive motor that will cause equipment to be out of action and has 60-day lead time might be class A, automobile batteries that are prone to disappearance and cost $50 each might be class B, and so on. Class should determine how often a part is cycle counted. No more than 10 percent of all SKUs should be class A, 30 percent class B, and the remaining 60 percent class C.

PUSH VERSUS PULL

Service parts business has traditionally been reactive to demands from the end users and hence classified as a "pull" channel of distribution. In recent years the concept of "push" has been advocated. This means to forcast what a particular user is going to require and to supply the correct number of parts at the right time without the user needing to ask for them.

The push method can have advantages when the manufacturing or central distribution facility is constrained as to space and cannot store the parts. As soon as a production run is manufactured, it will be allocated to the next level in the distribution chain and immediately shipped. If, for example, an order for 1000 motors is received at the national distribution center, then 200 might be allocated to each of five regional centers, which in turn allocate 20 to each of its ten districts, and these in turn send one to each local branch and distributor. Note that all shipments could be made most efficiently directly from the national distribution center, rather than requiring each level to break its bulk shipment and tranship to lower levels. Generally, instead of such uniform allocation, the parts are assigned on the basis of demand, historical usage, forcast requirements, and population supported.

SLOW MOVING ITEMS

Service parts have many low-use items. When products are planned and introduced, the initial provisioning may have indicated that parts should be stocked that are rarely used. The best answer is to establish guidelines as to how long a part should be retained before it is removed from inventory.

A field service organization with several levels of support, after consideration for insurance, long lead, and highly essential parts, should carry items in tech rep car trunks stocks only if they are used at least every three months. If not used in that time, the parts should be sent to the next higher level of supply. Let us assume that is a field branch. If the field branch has not used the part in six months, then it should be sent to a regional distribution center. If the center has not had a demand for the part in a year, it should be sent to a national consolidated stock location. It is vital that data be maintained on the last date parts are used and on turnover, so that parts can be removed from stock and sent to the next higher level.

In a plant maintenance organization or similar situation where there is no hierarchy of stockrooms, low-usage parts should be checked about every three months. If the parts are not highly essential, long lead and have not been used in the past quarter, then efforts should be made to reduce the stock to the number necessary for a single use. Review parts for possible elimination from stock as soon as possible, so some benefit may be derived from resale or other use. If you wait too long, then parts are of little value to anyone. Even if the parts have no value, they should be scrapped since they may be taxed and are at least taking up space in storage and in data systems.

LOCATION UNIQUENESS

When multiple plants or field locations exist, it would be very nice to stock all identically. That, however, probably would mean excess in some and shortages in others. The main factors that affect differences are customers/users, products supported, and travel time. For example, sophisticated reprographics equipment located in a printing plant in Kansas City is far from any similar units. To support that single reprographic system could require the same level of spares that five systems would require if they were within short distances of each other. If the plant or main support center for the equipment is nearby, then that provides a reserve source of parts not available at more distant locations. Therefore, installations close to the center could be stocked at a lower level and the center relied on for direct support.

If parts usage depends on customer applications, operator care and attention, or use magnitude, then those differences should be considered. Some parts will be used based on time, others on events, and others on operating conditions. Even worker-incentive programs can have a major impact on parts use. It is not unknown for a wrench to be dropped into equipment on Friday afternoon so the operator can have the rest of the shift off. That may be malicious equipment damage, but it still requires parts to repair. A story about location differences, alleged to be true, concerns Navy submarine periscopes. A sudden increase in damage to periscope tubes and the optics and electronics mounted on them was reported by one fleet. Upon investigation it was discovered that the submarines involved were engaged in escort duty and a procedure had been developed to use them to inspect the bottoms of large aircraft carriers for mines and other potentially damaging objects. A submarine, while submerged, would be brought near the large ship and the periscope used to view the underwater surfaces of the hull. Even a slight miscalculation could result in the periscope hitting the larger ship's hull. The result was usually a bent tube and shattered components. The embarrassment of having to return to port on the surface for repairs was only part of the anguish. The lost mission time and high dollar cost severely hampered several officers' careers. Here the reason for high parts usage was clearly operator failure. Corrective action was not to obtain more spares, but rather to direct the fleet to change the procedure and not engage in such damaging activities.

KITS

It makes good sense to combine related parts that are necessary for a specific job into a package usually called a kit. An automobile tune-up kit, for example, should have a complete set of spark plugs, ignition components, filters, and anything else necessary to do the job, including instructions. If the new part configuration may not be completely compatible with the removed part, then everything necessary to complete the installation should be included, even special tools. Preventive maintenance parts kits are an example where all components can be packaged to have everything easily available to the user. It should be possible to return used parts from a kit, possibly for credit, but at least for future use as individual replacement parts.

A parts kit should have a unique part number to identify the bill of materials. This allows users to order piece parts by their individual part stock numbers, or to order the complete kit under just one stock number.

Field service organizations that repair sophisticated products only infrequently, or that support third parties in maintenance of those products, will probably provide a complete kit of any parts that might be needed on a service call for that product. Printronix, for example, keeps kits at the Federal Express Parts Bank. When a service call is requested, the headquarters dispatcher notifies both the local service agent and the Parts Bank. The parts kit is shipped in time to arrive by the next morning, so that when the technician gets to the machine location the parts will be waiting. Any defective repairable parts are "red tagged" and put back in the kit, which is returned to the parts bank. A local service representative checks returning kits to replace any defective parts, arrange for repair of the defective parts, and assure that everything necessary is in the kit ready to go again. A similar system is used by many service organizations that keep kits at branch or district offices and arrange for pickup or shipment as necessary. Taxi, bus, courier services, and even a part-time parts delivery person can be effective in getting the best possible turnover from low-use but essential parts.

CONFIDENCE

Plant maintenance and field service personnel cannot be comfortable in their work unless they have an adequate supply of parts on hand. When angry customers complain loudly that the technician always seems to come the first time just to look and then has to come back again with the parts, the technician will do everything possible to obtain those parts and have them on the first call. The factory stockkeeper who has been upbraided for taking five days to get a cam that kept the main line out of production will try to have at least one spare cam secreted away in case of future need.

The key is confidence. If parts users have confidence that they can get a part quickly, then they will not keep many on hand. Confidence can be shattered quickly and takes a long time to regain. Those parts that are being retained "just in case" cost money to hold. Even if they cannot be sold for much, they are consuming valuable space and human energy that could be better devoted to having the right parts on hand. If logic and rational discussion do not prevail, then occasionally a manager must resort to authoritarian commands, and, of course, be willing to accept the consequences when a needed part occasionally "is just thrown out."

CANNIBALIZATION

Asking maintenance and service personnel if they have ever cannibalized equipment invariably brings laughter. It is a fact of life that should be addressed by specific guidelines. To say absolutely not is to be like an ostrich with its head buried in the sand, disregarding the potential benefits as well as the problems.

Removing parts from otherwise serviceable equipment should be avoided whenever possible. The cost of cannibalization is primarily double effort involved in removing a part for use someplace else, and then having to replace the same part when another is available. This, of course, exposes the equipment to potential damage during the activity. Too often the part is not reinstalled on the equipment. It is very easy to forget or purposely avoid doing that. Major problems often occur with equipment being returned for rebuilding or refurbishing. Field personnel are inclined to remove from those returning machines any parts for which they have use. This causes unexpected additional work and cost in the rebuilding process. Also, it devalues capital assets that actually may be owned by third parties and parts removal from them is technically a criminal act.

On the other hand, the benefits of being able to remove a part from the office demonstrator machine to appease an important customer or to remove a PCB from a disk that is already down for head repairs can be useful. The trade-off is usually against downtime. Be sure that any equipment from which parts are purposely removed has a tag or label that identifies the missing parts and who is responsible for getting them back on. Also, that person should make certain the parts are ordered and reinstalled on the equipment. Cannibalization should rarely be practiced, but in a difficult situation it could be the way out.

DISASTER STOCKS

Safety stocks have been already discussed as primarily to provide coverage for extremes of demand increases and delays in resupply. Disaster stocks are related, but should be considered as an insurance policy against physical disasters such as a fire, flood, explosion, or tornado that may wipe out a parts facility. Of course, if the facility is in a factory, then probably everything is wiped out so there is no need to have disaster stocks. However, a field service organization such as Amdahl, Mosler, or

Xerox should have contingency plans ready in case one parts support facility is knocked out. That preparation includes not only knowing from where the spare parts would be obtained, but also having backup data-processing facilities available, including customer lists and tech rep shipping addresses.

INVESTMENT STOCKS

When the price of gold began to climb, some parts managers said, "I'm going to buy twice my normal quantity of printed circuit boards because with all those gold contacts their price will be inflating and I'll save the company a lot of money." When the worldwide shortage of petroleum occurred and plastics became scarce, the same people bought large quantities of plastic cams and covers. Today they look very sad when those situations are mentioned. There were many circuit boards that were scrapped as obsolete and covers that were never used. Generally, investing in extra parts to make money is not advisable. Chances are that carrying costs in the area of 35 percent a year are far more important than any inflationary value of the parts.

WARRANTIES

Parts should normally be warranted against defects for 30 or 90 days. Unlike warranties on equipment, warranties on parts are rarely a significant factor in the purchaser's decision-making process. If the choice is between a new or a rebuilt starter, the 90-day warranty may sway the customer to accept a perceived (but probably erroneous) risk associated with the rebuilt part.

A challenge is how to calculate the warranty time on parts. In most cases the part is purchased to sit on the shelf for months before it is installed in any equipment. Does the warranty begin when the manufacturer ships the part? Does it begin when the part is sold to the end user? Or does it begin when the part is put into use? If the last is desired, then how is it known when the part is placed in service?

The fact is that parts warranties are largely neglected. Most plants that install parts that fail during the warranty period either are not aware that the part was covered by warranty, do not have a record system to track the failure and get the parts back to the vendor for repair or replacement, are too busy to handle it, or just do not care enough to make the effort. The size of the using organization and the volume of failed parts are the main determinants. If expensive motors begin to fail under warranty, they will certainly be the subject of a special program to determine the faults and return them to the vendor for replacement or financial reimbursement. From the vendor's perspective, a good customer who wants satisfaction will probably get free replacement parts even though the failure may be slightly over warranty. If the joke about a part failing on the 91st day should come true, most service organizations would replace that part free as a good-will gesture.

Controlling the return of parts under warranty and avoiding abuses is usually done by requiring return of those parts or a significant component of them. If, for example, a power supply is to be returned and it is not repairable, the supplier probably will request that the electric leads be cut from it and returned. The rest of the power supply can be scrapped as of litle value and the supplier knows by the return of the wires that the supply is beyond use to anyone. Other organizations require the claimant to retain the parts for 30 days in case engineering needs them returned for technical evaluation. Multitiered parts distribution organizations working through distributors and dealers often have their field managers occasionally check the collection of parts under warranty, to make sure they are legitimate.

A stimulating question is whether parts warranties should include the labor costs for installation of the defective, covered part. The customer would certainly like that to be so. An organization that provides contract service or leases including full support must replace the part and then try to get expenses back from the vendor of the defective material—whether its own manufacturing facility or an outside supplier. From the supplier's perspective, warranties are primarily a customer relations tool. If high costs are involved, a notice of installation may be necessary for control. Testing to assure the failure was due to defective workmanship or materials can avoid unnecessary expenses, but borderline situations generally should be decided in favor of the customer.

LIABILITY

By selling a part or by installing it in equipment, the manufacturer and supplier and service organization represent legally that the part is fit for the anticipated use. Formal procedures should be practiced to assure that the parts as received by the end user meet specifications and are properly installed, if applicable.

PARTS USED AS DIAGNOSTIC TOOLS

Many parts are stocked as much for their use in diagnosing problems as for replacements. If an electronic problem cannot be diagnosed by test equipment, the technician probably will pull the most likely defective board and replace it with a new one. If that was not the problem, then the next board will be replaced, and so on until the equipment operates properly.

The calculations for deciding whether separate diagnostics should be designed into equipment or tools requires knowledge of what the dollar value of spares used as replacement parts is versus the dollar value of those parts used as tools. If every technician has a spare circuit board for use as a tool, that cost can be very high. For example, if each of those boards is worth $400, then 500 technicians times $400 each means $200,000. Certainly, even a portion of that would be adequate budget to design diagnostics and reduce the number of parts that are carried to function as tools. Of course, if the parts are going to be carried anyway as replacement spares, then they can serve a dual purpose. The use of parts as tools is often overlooked in the decision to share kits among many technicians. From the numbers one may calculate that, say, two spares will support the 100 machines serviced by five technicians. This may be fine for the failure-rate calculations but may not be accurate for the true mean time between replacement or the need for those parts as diagnostic tools. Remember also that parts may not be of 100 percent perfect quality themselves, so on occasion a bad part being used as a diagnostic tool can cause considerable anxiety.

SUCCESS FACTORS

Managing service parts is a challenging job. There are many factors that contribute to success. A large electronics manufacturer's service parts organization has found the list shown in Table 19-2 to be most important.

Table 19-2   Success Factors—Service Parts Management Central Inventory

1. Simple, understandable forecasts; different forecasts for different parts and demand types
2. High inventory and order accuracy
3. Exception condition reporting at the analyst level; exception condition management reports
4. Minimum stock levels for low-demand spare parts; controls on economic order quantities
5. Proper handling of engineering change orders; prompt disposition of superseded material
6. Automated link of forecast data with manufacturer; automated purchase order link with manufacturer
7. Monitor inventory obsolescence rate; budget and carry out scrap plan
8. Measure first-pass fill rate and order aging; measure in-stock rate by class
9. Manage items based on hit rate as well as class; manage customer orders
10. Recycle repairable parts promptly; ensure highest quality standards are used for spares testing

KISS

"Keep it short and simple" is good guidance to parts management.

Index